Dynamics

an introduction for civil and structural engineers

ICE design and practice guides

One of the major aims of the Institution of Civil Engineers is to provide its members with opportunities for continuing professional development. One method by which the Institution is achieving this is the production of design and practice guides on topics relevant to the professional activities of its members. The purpose of the guides is to provide an introduction to the main principles and important aspects of the particular subject, and to offer guidance as to appropriate sources of more detailed information.

The Institution has targeted as its principal audience practising civil engineers who are not expert in or familiar with the subject matter. This group includes recently graduated engineers who are undergoing their professional training and more experienced engineers whose work experience has not previously led them into the subject area in any detail. Those professionals who are more familiar with the subject may also find the guides of value as a handy overview or summary of the principal issues.

Where appropriate, the guides will feature checklists to be used as an *aide-mémoire* on major aspects of the subject and will provide, through references and bibliographies, guidance on authoritative, relevant and up-to-date published documents to which reference should be made for reliable and more detailed guidance.

ICE design and practice guide

Dynamics

an introduction for civil and structural engineers

J.R. Maguire and T.A. Wyatt

  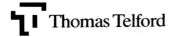

Published by Thomas Telford Publishing, Thomas Telford Ltd, 1 Heron Quay, London E14 4JD
http://www.t-telford.co.uk

First published 1999
Reprinted, with corrections 2000

Distributors for Thomas Telford books are
USA: ASCE Press, 1801 Alexander Bell Drive, Reston, VA 20191-4400, USA
Japan: Maruzen Co. Ltd, Book Department, 3–10 Nihonbashi 2-chome, Chuo-ku, Tokyo 103
Australia: DA Books and Journals, 648 Whitehorse Road, Mitcham 3132, Victoria

A catalogue record for this book is available from the British Library

Classification
Availability: Unrestricted
Content: Recommendations based on current practice
Status: Refereed
User: Practising civil engineers and designers

ISBN: 0 7277 2522 X

© Institution of Civil Engineers, 1999

All rights, including translation reserved. Except for fair copying, no part of this publication may be reproduced, stored in a retrieval system or transmitted in any form or by any means, electronic, mechanical, photocopying or otherwise, without the prior written permission of the Books Publisher, Publishing Division, Thomas Telford Ltd, Thomas Telford House, 1 Heron Quay, London E14 4JD.

Every effort has been made to ensure that the statements made and the opinions expressed in this publication provide a safe and accurate guide; however, no liability or responsibility of any kind can be accepted in this respect by the publishers or the authors.

Typeset by Gray Publishing, Tunbridge Wells, Kent
Printed in Great Britain by the Cromwell Press

Foreword

Dynamics is a far more important subject to civil and structural engineers than it used to be, because structures have become lighter, members more slender and, for buildings especially, cladding has changed from brick and masonry to steel and composite materials in much of modern architecture. These changes have increased amplitudes of vibration and moved frequencies of structures into bands which are both more awkward to deal with as well as being more easily perceived by users.

This guide provides an introduction for practising civil and structural engineers seeking to find out more about the subject of dynamics. It is aimed primarily at those approaching chartered status, whatever their age, and will educate practising engineers in the main principles and important aspects of the subject. It provides through the references guidance on authoritative, relevant and up-to-date published documents which practising engineers should refer to for more detailed and reliable guidance. We believe we have provided all the information necessary to satisfy these objectives in a concise manner.

It is important to stress that dynamics is a complex and evolving field. This guide is intended as an introductory reference document and should not be expected to provide detailed information, or solutions to, all dynamics problems. If in doubt the reader should seek the assistance of a recognized dynamics specialist.

Acknowledgements

This guide has been mainly prepared by Dr John Maguire of Lloyd's Register and Professor Tom Wyatt of Imperial College, together with other members of SECED (Society for Earthquake and Civil Engineering Dynamics) and WES (Wind Engineering Society), both learned societies within the Institution of Civil Engineers. Their contributions have been primarily as individuals and should not be construed necessarily as the views of their sponsoring organizations.

The contributions and assistance of the following in reviewing and contributing to the text are gratefully acknowledged:

Edmund Booth	Independent Consulting Engineer
Professor Norman Jones	Impact Research Centre, University of Liverpool
Dr Brian Ellis	Building Research Establishment
Dr Bryan Skipp	Independent Consulting Engineer
Professor David Key	CEP Research
Dr Peter Merriman	British Nuclear Fuels Ltd
Dr Alan Watson	University of Sheffield
Dr Maurice Petyt	ISVR, Southampton

This guide also draws significantly on work produced by Dennis Hitchings, Imperial College, and Professor Arthur Bolton of Heriot-Watt University, and their material is acknowledged accordingly.

Production of this guide has been the responsibility of the Institution of Civil Engineers and Thomas Telford Limited.

Contents

1.	Introduction	1
2.	**Basic dynamics theory**	**4**
	Input (loading)	4
	The dynamic system	5
	Equation of motion	6
	Modes of vibration	7
	Output (response)	9
3.	**Design for dynamic loading**	**12**
	Natural frequencies	12
	Damping and dampers	14
	Acceptance criteria: ultimate limit state	15
	Acceptance criteria: other limit states	16
	Initial design	19
4.	**Specific dynamic loadings**	**21**
	Wind-induced vibrations of structures	21
	Earthquake-induced vibration of structures	30
	Vibration induced by people	39
	Blast effects	43
	Machinery	48
	Ground transmitted vibration	51
	Impact	54
	Hydrodynamic loading: wave- and current-induced vibrations	57
5.	**Summary of selected design issues**	**60**
	From Chapter 1 – introduction	60
	From Chapter 2 – basic dynamics theory	60
	From Chapter 3 – design for dynamic loading – general	61
	From Chapter 3 – design for dynamic loading – initial design	61
	From Chapter 4 – specific dynamic loadings	61
	Illustrative example 1: response of buildings to a gusty wind	63
	Illustrative example 2: response of a building to an earthquake	68
	Illustrative example 3: floor subject to rhythmic activity	72
Classified selected vibration standards		**74**
References combined with selected codes and standards		**75**

1. Introduction

What is dynamics? Why do I have to know about it? How should I accommodate dynamics in the civil engineering design and design checking process? These questions lie at the heart of this guide, and in attempting to address them, the starting point is statics. Any engineering designer is familiar with the concept of a structure to fulfil the set functional requirements which is checked to ensure that it can robustly withstand the action of one or more sets of static loads. Only rarely, however, are the forces acting on a real structure wholly unchanging with time and thus truly static.

The principal distinctive feature of a dynamic analysis is consideration of inertial effects, i.e. inclusion in the analysis of terms of the form (*mass times acceleration*). Ideally, one could solve the resulting equations of motion and use the results for design checking in the same way as with conventional static checks. A full knowledge of the displacements (deformation) of the structure can be fed back to check the internal stresses, but can also be applied to serviceability checks on displacements or on motion-perception criteria or to fatigue cycle counting.

In practice, such a rigorous extension rarely is practicable. A rigorous dynamic analysis inevitably requires much more computation than a static analysis; this is especially noticeable when a non-linear structural response characteristic is involved. Even if the computational effort was practicable, many engineers may lack the experience and insight into dynamic design checking to be sure that they are correctly assimilating their accumulated static experience and insight. Fortunately, in many cases it is possible to give guidance by which the designer can decide whether:

— dynamics is of marginal importance and is safely covered by load factors in conventional checks; or
— dynamics is significant and specific practical design-office procedures should be applied; or
— dynamics is crucial and specialist measures and/or advice should be considered.

One especially dangerous phenomenon is resonance, resulting from synchronism between a pronounced periodicity in the loading process and a natural vibration frequency of the structure. Resonance can lead to gross magnification of response (one-hundred-fold is not impossible), with potential overstressing or rapid fatigue failure. Every effort should therefore be made to avoid resonance, although this is not always economically practicable; as discussed later, an economic design to a predetermined structural form leads to a broadly predetermined natural frequency.

Where resonance cannot be avoided, it may be possible to reduce the magnitude of the excitation, for example by vibration-control mounting of machinery or by modification of the aerodynamic shape of the structure. Further examples of countermeasures include isolation from ground-borne vibration, and enhancement of natural damping to reduce the dynamic magnification. Similar countermeasures may be applied in many of the cases where a significant dynamic response at a natural vibration frequency arises although the input is seemingly random; such cases include earthquake, wind gust, wave and traffic effects.

A somewhat different behaviour emerges in consideration of major impulsive 'accidental' loads; collisions with bridge piers, offshore structures, and so on. The requirement is a structure that can safely absorb energy, with emphasis on non-linear structural behaviour rather than dynamic theory. Lack of space does not permit substantial treatment in this guide, although a chapter is devoted to it together with substantial references.

Natural frequency, mass and energy dissipation capability are thus all-important. Dynamic response has become more important as a result of trends in all three properties. Natural frequencies have tended to decrease in recent decades because working stresses have increased (as a consequence of both higher-strength materials and more refined design) without commensurate increases in Young's modulus. Response to wind gusts is thereby considerably increased because not only does this imply a shift towards the peak ordinates of the wind-gust spectrum but the 'quasi-resonant' gusts are larger, as described later. Critical wind speeds for aerodynamic instability problems are generally reduced. Similar trends apply to so-called 'fixed' (as opposed to 'compliant') structures in water. Excitation by traffic (people or vehicles) is also commonly exacerbated, bringing resonance into play either with a more severe Fourier component for a given span of floor, or at a shorter span of a bridge. Earthquake effects may or may not be adversely affected by reduction of structural natural frequency. Reduction of mass is generally adverse where the acceptability criterion is a question of human subjective reaction; other things being equal, response acceleration is proportional to excitation divided by the mass.

In many cases, energy dissipation or damping plays a key role. Within the range of response that would be acceptable for any excitation of cumulative duration measured in hours rather than seconds, the natural damping of structures is most often reliant largely on frictional dissipation of energy. Over recent decades welding or friction-grip bolting has replaced 'black' bolting; prestressing has been introduced; building partitioning has been reduced; building cladding has come to be carried on 'engineered' resilient mountings to reduce its susceptibility to frame deformation; the number of expansion-bearings and movement joints has been reduced by continuous construction: all the foregoing have had the effect that frictional dissipation of energy has been reduced. It is notable that many laudable design advances are actually adverse from the viewpoint of dynamic sensitivity.

There are very diverse potential acceptance criteria for dynamic response. Earthquake is again an extreme case with the dominant criterion of prevention of complete collapse, because the balance of cost against probability of occurrence marginalizes the importance of lower levels of damage. For the other dynamic problems, analysis is usually linear, carrying the implication that the stresses resulting from dynamic response can be superposed on the static equilibrium solution (a slightly more sophisticated probabilistic approach is taken in wind-gust problems but it is still implicitly 'elastic'), even if the structural capacity estimate is based on ultimate resistance. At first sight this is conservative, but it is not necessarily so.

In many cases, large numbers of response cycles may accumulate over the life of the structure, and the linear solution is then directly appropriate for a fatigue check. Closed-form solutions for fatigue cycle counting based on the Palmgren–Miner law are available in some cases. Some of the design trends previously mentioned (higher general stress levels, welded fabrication, ultimate-strength design permitting relatively unfavourable distribution in the elastic range) increase the impact of fatigue on the designer.

Finally, in many cases the question at issue is human response to the perception of motion. This is an exceptionally difficult question because it is generally not a physiological question of danger or of ability to perform physical tasks, but a psychological question of feelings of security or intrusion on concentration. The conclusions become very uncertain due to variability between people, variability according to activity and other features of the environment, variability according to expectation and familiarity. Significant improvement of a case where the predicted subjective response is marginally unfavourable is commonly expensive, and the degree to which explicit mandatory criteria would be appropriate is still controversial. It is nevertheless important that the designer appreciates the problem and can present a clear assessment. Specialist design intervention in cases of subjective response are, however, well established; for example, the isolation of buildings from ground-transmitted noise and vibration, such as from railways.

This guide is therefore structured as follows:

— Chapter 2 gives the basic theory underlying dynamics;
— Chapter 3 considers what are the acceptance criteria in cases where dynamic loading is significant in design. The final section of this chapter in particular summarizes some important initial design considerations;
— Chapter 4 considers a broad range of dynamic loadings which may be significant in many design situations;
— Chapter 5 provides a summary of selected design issues as an *aide-mémoire* to designers who wish to revisit parts of the guide which they may have noted as particularly useful;
— the guide concludes with a number of illustrative examples, a classification of vibration standards, and a set of references which includes relevant codes and standards.

2. Basic dynamics theory

Input (loading)

Dynamic loading can be classified into a number of types depending on the nature of the timewise variation. The classification used in this guide is as follows:

(a) periodic or harmonic (load amplitude repeats itself regularly many times);
(b) transient (load varies with time but does not repeat itself continuously);
(c) stationary random (load not known precisely but statistical properties vary only very slowly);
(d) non-stationary random (as for stationary but varying statistical properties).

These four main loading types are illustrated in Figure 1, as plots of loads against time. This representation is known as a time-domain representation. The same loading types could also be represented as a plot of amplitude against frequency, which is known as a frequency-domain representation.

A frequency-domain equivalent of Figure 1 is illustrated in Figure 2. Note that a 'full' frequency-domain representation comprises both an amplitude–frequency and a phase–frequency plot – in Figure 2 only the former is shown. The usefulness of a frequency-domain plot is that it can often highlight the most significant characteristics of the loading which may be obscure in the time-domain.

Whereas the vertical ordinates of Figures 1(a) and 2(a) have the conventional direct interpretation as forces, it is more common for random excitation (Figures 1(c) and 2(b)) to present 'power spectra', which have the units (dimensions) of the square of the basic input variable per unit of frequency; for example, $(m/s)^2/Hz$ for a wind speed spectrum. The power spectrum takes its rigorous definition through the Fourier integral transform, which is beyond the scope of this book (see Newland, 1975). Nevertheless, much insight can be gained from an heuristic interpretation as follows.

The area under the spectral curve is equal to the variance (mean-square departure from the concurrent mean value) of the respective 'parent' process. Thus, if the spectrum of response can be evaluated, the variance and thus the RMS (root mean square) value can be determined. A simplified interpretation of the spectrum showing how the variance is apportioned to constant-amplitude sinusoidal components can be useful. As the spectral ordinates relate to the square of the parent unit, however, the 'transfer functions' by which load is converted to response follow the squares of the conventional deterministic relationships. In particular, this gives a direct picture of the importance of the natural frequency and 'quasi-resonant' response (see pages 9 and 21).

Basic dynamics theory

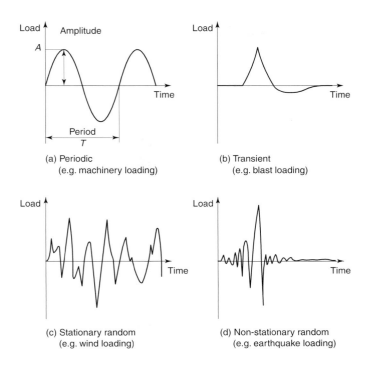

Figure 1 Examples of dynamic loading (time-domain representation)

Acceptance criteria, such as the maximum value expected to occur in a given duration of exposure, or fatigue or subjective comfort responses, can then be checked by reference to the RMS value, with application of an appropriate numerical multiplier.

The dynamic system

Many civil engineering structures may be treated as linear systems, with the implication that the principle of superposition is valid, or as non-linear. Despite the widespread recognition of non-linearity in quasi-static design, at least for the ultimate-resistance limit state, linear analysis remains common in dynamics. A crucial reason for this is to facilitate evaluation by *modal analysis* (see page 6). Linear analysis is clearly appropriate for many dynamic criteria (fatigue, comfort), and in other cases (many wind, wave and traffic problems) the dynamic response is largely within the linear limit. The principal cases for non-linear analysis are seismic loading (see page 30) and 'accidental' impact loadings (see page 54).

Any system must have certain characteristics before it will vibrate. It must have a stable position of equilibrium so that, if it is disturbed for any reason, it tries to regain

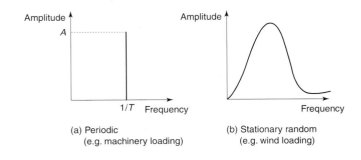

Figure 2 Examples of dynamic loading (frequency-domain representation)

this stable position. The restoring force trying to regain the equilibrium position is called the stiffness force. For structural vibrations the stiffness force is proportional to the displacement of the structure and the coefficient of proportionality is called the stiffness (k, see Figure 3) of the structure. The stiffness stores potential energy which, for structural systems, is usually the strain energy within the structure, but gravitational or buoyancy effects may also make a significant contribution. For vibrations to occur the structure must also possess mass; it is the momentum of the mass that causes a vibrating structure repeatedly to overshoot its equilibrium position. It is often useful to think of an inertia force, acting on the structure, equal (but opposite) to the product of mass and acceleration. Motion of the mass also implies kinetic energy; vibrations are the physical manifestation of the interchange between potential and kinetic energies.

The stiffness force is in phase with the vibrational displacement, and thus no net work is done in a cycle of vibration. All practical systems must contain some energy-dissipating mechanism; the system is then said to be damped. If no external forces are applied to input energy into the system the damping will cause the amplitude of the displacements to die away with time. A common idealization used for damping is to assume that the damping force is proportional to the velocity of the structure. In this case the damping is said to be viscous and the constant of proportionality between the damping force and the velocity is the viscous damping coefficient (c, see Figure 3).

Equation of motion

The equation of motion defining the dynamic behaviour of the structure is the equation of equilibrium between the inertia force, damping force and stiffness force together with the externally applied force. This is of the form:

inertia force + damping force + stiffness force = external force

or, in algebraic form:

$$m\ddot{y} + c\dot{y} + ky = f(t) \qquad (2.1)$$

The linear dynamic behaviour of almost any structure is controlled by an equation of this form. The dynamic response is found by solving this equation of motion. Before these can be solved the structural boundary conditions must be specified to define how a structure is supported. In many cases it is also necessary to define a set of initial conditions for the displacement and velocity at time equal zero in order to be able to solve the equations.

In a structure of practical complexity the displacements will be characterized by values at selected locations, the degrees of freedom (DOF). These must be so chosen as to enable a sufficiently accurate estimate of the kinetic energy to be made. If a general-purpose computer package is used, the DOF may simply be the nodal displacements as

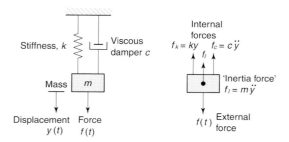

Figure 3 The basic dynamic system

conventionally used, but care must be taken to ensure that this gives sufficient representation of the motion of mass at intermediate points, and it may be necessary to introduce additional nodes for this purpose. On the other hand, many rotational component DOF in a conventional model have very little effect on dynamics. Many *ad hoc* dynamic procedures eliminate (or ignore) these DOF to reduce the order of the problem. If the stiffness matrix is called **K**, the mass matrix called **M**, and the damping matrix called **C**, then the damped equation of motion for any linear structural in matrix form is:

$$\mathbf{M\ddot{y} + C\dot{y} + Ky = F} \tag{2.2}$$

The mass is often simply lumped at the nodes, giving a diagonal mass matrix. A more complex procedure which significantly reduces the need for additional nodes introduces the so-called consistent mass matrix; see Clough and Penzien (1975).

If damping is excluded and no external forces are applied then the homogeneous undamped equation of motion is:

$$\mathbf{M\ddot{y} + Ky = 0} \tag{2.3}$$

This has a solution in the form of simple harmonic (periodic) motion:

$$\mathbf{y = \tilde{y}}\sin \varpi t \quad \text{and} \quad \mathbf{\ddot{y}} = -\varpi^2 \mathbf{\tilde{y}} \sin \varpi t$$

Substituting these into the equation of motion gives:

$$\mathbf{K\tilde{y}}_0 = \varpi^2 \mathbf{M\tilde{y}}_0 \tag{2.4}$$

This is known as the eigenvalue problem, where ϖ^2 is the eigenvalue and $\mathbf{y}_0$ is the eigenvector. ϖ is the natural frequency in radians per second so that the eigenvalue is the square of the natural frequency. $n = \varpi/2\pi$ is the corresponding natural frequency in cycles per second (Hz).

Modes of vibration

If the system has r DOF, equation (2.4) will have r solutions, representing the normal modes of vibration of the system. In each mode, a free vibration consists of harmonic motion of all points in phase, at the frequency n_i (or the circular frequency ϖ_i), in which $y_j = Y_i \Phi_i \sin \varpi_i t$, in which Y_i is the modal generalized amplitude and the eigenvector Φ_i (comprising elements ϕ_{ij} for each DOF j) is the mode shape vector (Figure 4).

The normal modes are uncoupled, i.e. external or internal forces associated with any one mode do not affect any other mode. For each mode a generalized mass M_i and generalized stiffness K_i can be evaluated from the eigenvalue solution, i.e. $M_i = \Phi_i M \Phi_i^T = \sum m_i \phi_{ji}^2$ (the sum being taken over all DOF j) and $K_i = \varpi_i^2 M_i$. Any

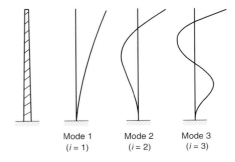

Figure 4 The first three modes of a cantilever beam

Mode 1 ($i = 1$)   Mode 2 ($i = 2$)   Mode 3 ($i = 3$)

external load vector **F** (comprising elements f_j) can similarly be resolved into modal components as scalar quantities $F_i = \sum f_j \phi_{ji}$. The uncoupled property of the modes then means that the full problem is reduced to superposition of the modal responses, each obtained using the simple single degree of freedom solution for the given $F_i(t)$, K_i and ϖ_i. The modal generalized stiffness increases with mode order (order of natural frequencies) and many practical excitations thus call only for the first, or the first few, modal solutions.

To retain the uncoupled property when damping is included, it is strictly necessary to impose conditions on the relative values of the elements making up the damping matrix **C**. However, for most structural systems the damping force is small compared to either the inertia or the stiffness force. For this reason it is often only necessary to model an appropriate net dissipative action for the various modes of oscillation, without defining exactly how this operates within the structure. Such measures are the logarithmic decrement (Figure 5) ('log dec') and the critical damping ratio. For small damping, a log dec δ implies that in a free decay of vibration the amplitude of each cycle is $1 - \delta$ times the amplitude of the immediately preceding cycle. The critical damping ratio is ξ related to δ by $\xi = \delta/2\pi$. For an explanation of more complex representations of damping (e.g. material, hysteric and friction) the reader is referred to Bolton (1994, pages 10–16), Hitchings (1992, pages 170–196) or Irvine (1986, pages 40–45).

For each mode, given an instantaneous value of the modal displacement Y_i (say), the contributions to the displacement in DOF j is $Y_i \phi_{ji}$, where ϕ_{ji} is the 'j' element of Φ_i. A similar procedure gives the contribution to the internal stress, which is $Y_i \beta_{ji}$, where β_{ji} (for any stress identified by suffix j) is the value of stress associated with unit displacement in mode 'i'. These modal stress factors can often be drawn directly from a computer-based dynamics package, with due caution that accuracy of stress will be more affected than displacements or natural frequencies by, for example, approximations arising in the discretization of the system. This can be augmented or circumvented by using the modal shape vectors to create the inertia load-case $f = m_i \varpi^2 \phi_{ji}$ (Figure 3) which is returned to the designer's static analysis.

The foregoing results for a discretized model of the structure have direct equivalents in a continuous modal, the mode shapes being continuous functions $\phi_i(s)$ of the location coordinates. The modal generalized properties then take the integral equivalents of the summation expressions given above, i.e. taking $m(s)\mathrm{d}s$ in place of m_i, where $m(s)$ is the mass per unit length. Extension to two or three dimensions is also straightforward.

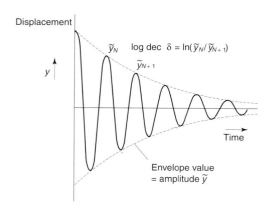

Figure 5 Free decay of vibration

The solution for the first mode of a simply supported uniform beam of span L is

$$\phi(s) = \sin \pi s/L, \quad \varpi_1^2 = \pi^4 EI/mL^4 \qquad (2.5)$$

in which EI is the flexural rigidity and m is the mass per unit length. The first mode frequency expressed in Hz is thus

$$n_1 = \frac{\pi}{2}\sqrt{\frac{EI}{mL^4}} \qquad (2.6)$$

The corresponding analysis for a uniform cantilever leads to a transcedential equation with solution

$$n_1 = 0.56\sqrt{\frac{EI}{mL^4}} \qquad (2.7)$$

The mode shape is approximately $\phi(s) = s^2(3-s)/2$.

Modal analysis is thus an extremely powerful tool, and is adopted almost universally. The principal shortcoming emerges from the designer's common wish to examine maximum values of response over time. Although superposition applies, clearly the maximum combined value is less than the sum of the individual maxima, which occur at different instants. A statistical combination procedure is commonly used for response to random excitation, but a laborious trawl through the time history for each output quantity is otherwise necessary.

For structures with moderate geometric or elastic material non-linearity, linear procedures can often be applied, using the stiffness defined by the tangent to the load–deflection curve at the point of state equilibrium under mean loading; suspension bridges and guyed masts are examples. For structures with moderate elastoplastic non-linearity (for example, complex structures where dynamic excursions beyond the linear-elastic limit are localized and limited in extent and duration), linear procedures are again commonly useful, with the energy dissipation by yielding having the effect of enhanced damping. The direct time-domain solution (see page 11) may be the only realistic procedure if larger plastic excursions are acceptable. Caution is also necessary that non-linearity can lead to patterns of behaviour not at all revealed by linear analysis; hanging cables (such as mast stay ropes) and pendulum-related systems (including 'tension leg' buoyant platforms) can in such cases suffer serious dynamic amplification of periodic forces of twice the structural natural frequency at which the oscillation will arise.

Output (response) Modal analysis provides the key to linear analysis by synthesizing the response from simple single DOF (SDOF) results. There are three principal cases:

— steady-state response to harmonic (sinusoidal) excitation, including spectral (Fourier integral transform) random loading;
— response to transient impulsive or step-change forces;
— complex deterministic force time-histories.

Harmonic excitation The steady-state response amplitude $\tilde{y}$ resulting from harmonic force amplitude $\tilde{f}$ at frequency n acting on an SDOF system of mass m and stiffness $k = m\varpi^2$ (where $n_1 = \varpi/2\pi$ is the natural frequency) is given by $\tilde{y} = H\tilde{f}/k$. H is the dynamic amplification factor (DAF), or frequency response function, given by

$$H = \frac{1}{\sqrt{\left(1 - \frac{n^2}{n_1^2}\right)^2 + \left(\frac{\delta n}{\pi n_1}\right)^2}} \quad (2.8)$$

in which δ is the damping expressed as logarithmic decrement. This function is shown in Figure 6. The peak value, at resonance ($n = n_1$) is π/δ; thus in the event of (say) $\delta = 0.05$ (cf. page 14), the peak value is about 60. The crucial importance of damping is clear. Modal generalized values may be substituted in the above.

Under random loading, the frequency domain representation (Figure 2(b)) will commonly indicate that a natural frequency will lie within the range of frequencies of the excitation. In this case the structure will pick up the frequency components close to the natural frequency in the same way that a tuned circuit picks up a radio signal, and the response will generally contain a clearly-recognizable natural-frequency vibration (but not at a constant amplitude). This may be called the 'quasi-resonant' or 'narrow band' (because it actually comprises a narrow band of frequencies) part of the response.

The corresponding spectral transfer function is H^2/k^2 and is thus even sharper peaked. The key output quantity in the spectral analysis is the variance, given by the area under the response spectrum. An excellent approximation to the area contributed by the narrow band is given by multiplying the peak value by the effective bandwidth $n_i\delta/2$. The RMS value of response is in this case inversely proportional to the square root of the damping.

Transient, impulsive or step-change excitation

For transient, as opposed to periodic, loading the equivalent DAF is shown in Figure 7. The key symbols used are:

— duration of impulse = t_p
— natural period of structure = T_1 ($T_1 = 1/n_1$).

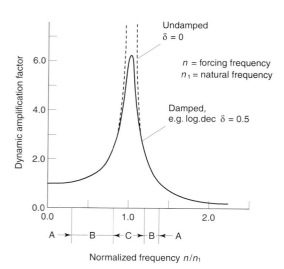

Figure 6 Dynamic amplification factor – harmonic loading (labels A–C are discussed in the text)

Basic dynamics theory

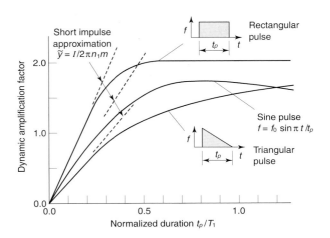

Figure 7 Dynamic amplification factor – transient loading

Further examples of transient excitation are evaluated in Thomson (1983) and Irvine (1986). When the duration is short, $t_p \ll T_1$ ($t_p < 0.2T_1$ is generally acceptable) the transient approaches a true impulse, i.e. inducing a change of momentum equal to the impulse defined by $I = \int f(t)\,dt$. The response is thus a free vibration with starting condition $\dot{y} = I/m$, giving peak response $\tilde{y} = I/2\pi n_1 m$.

Complex deterministic force time-histories

For cases other than the above, the equations of motion can be solved numerically using some form of step-by-step integration. There are many variations of these methods, some of which are conditionally stable in that they require the integration time step to be less than some value which is a function of the highest eigenvalue of the equations of motion. Other methods are unconditionally stable in that any time step length can be used. However, the results become progressively more inaccurate as the size of the time step is increased. These step-by-step integration methods are the only general solution techniques available for solving non-linear problems.

Problems of random excitation involving non-linear response, or where the duration of the excitation is insufficient to justify reliance on the steady-state solution, can also be studied by step-by-step integration with computer-generated input time-histories. These can be repeated as necessary to obtain robust statistical output. Economical and reliable step-by-step analyses generally require specialist participation – for further details see Hitchings (1992).

3. Design for dynamic loading

Natural frequencies

Although formal handling of the eigenvalue problem (see page 6) is the basis of computer-based evaluation of natural frequencies, much simpler approaches (beneficially supported by computer evaluation of static deflection) are usually practicable for the lowest mode. The Rayleigh method in its simplest form (e.g. Clough, 1975) is based on evaluation of the deflection (y_w say) caused by a force equal in magnitude to the self-weight of the structure but in the direction of deflection of the mode in question; a surprisingly good estimate of the lowest natural frequency is given by

$$\varpi_1 = 2\pi n_1 = \sqrt{(g/y_w)} \qquad (3.1)$$

An extension using an intelligently-selected trial load (f_t, say) greatly extends the scope of this approach to cover very many practical cases. If the structure is treated as having distributed mass, $m(s)$ per unit length where s is the local location coordinate, the trial load is postulated as a function varying in proportion to the product $m(s)\phi(s)$, where $\phi(s)$ is a guess at the mode shape satisfying the support, boundary and/or continuity conditions. The static deflection under this load is evaluated (y_t) and equation of the resulting estimates of the maximum values of the kinetic energy and the strain energy gives

$$\varpi_1^2 = (2\pi n_1)^2 = \int f_t(s)y_t(s)\,\mathrm{d}s / \int m(s)y_t^2(s)\,\mathrm{d}s \qquad (3.2)$$

The simplicity of this dependence on the mass and the basic deflection calculation has the effect that the designer has little power to alter the natural frequency, given an economic design to a conventional structural form. The lowest natural frequency generally follows a defined relationship to the size of the structure.

The economic design–natural frequency relationship is strongest where the form is most directly related to a strength requirement. Most steel bridges have natural frequencies for vertical oscillation within 20% of the values indicated in Figure 8.

For steel lattice towers the natural frequency is strongly related to the height H, base width B and the ratio of the factored design wind load to the weight (R_1, say, averaged

Design for dynamic loading

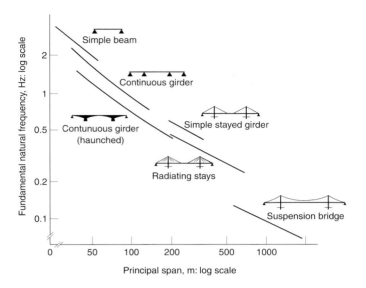

Figure 8 Fundamental natural frequency of steel bridges

over the top quarter of the tower; cf. BS8100), typically

$$n_1 = \sqrt{gR_1 B/H^2} \qquad (3.3)$$

R_1 is commonly between 0.5 and 1.2 for tubular towers, with a trend to higher values with reducing values of H, and typically 60% larger for angle construction. Thus for a typical tubular flare tower, $R_1 = 0.9$, $H = 70$ m and $H/B = 5.5$, $n_1 = 1.22$ Hz. The numerical constant is lower in the event of a strongly Eiffelized profile.

For concrete chimneys the natural frequency can be estimated from the diameters D_u, D_ℓ and shell thickness t_u, t_ℓ at the upper (u) and lower (ℓ) quarter-points of the height (respectively) and the ratio of the mass per unit height of non-structural material (linings, etc.) to that of the shell at the upper quarter-point, R_2 (say)

$$n_1 = \frac{D_\ell}{5H} \sqrt{\frac{E D_\ell t_\ell}{(1 + R_2)\rho H^2 D_u t_u}} \qquad (3.4)$$

Thus for height $H = 70$ m, $H/D_\ell = 12$, $D_u t_u = 0.6\ D_\ell t_\ell$, concrete with a dynamic Young's modulus $E = 32$ kN/mm^2 and density $\rho = 2.4$ t/m^3 and with lining $R_2 = 0.25$, the lowest natural frequency is $n_1 = 1.0$ Hz.

For buildings the scatter relative to the predictions that can be inferred from economic design for the normal gravitational or wind loading is unfortunately much more severe. The structural forms are more diverse and the disparity of stiffness between the designer's model and reality may be considerable (greater stiffness at connections, contribution of 'non-structural' claddings and partitions). The best that can be offered are global rules-of-thumb, successors to the long-established $n_1 = 10/N$, where N is the number of storeys. An extensive study by the Building Research Establishment (BRE; Ellis and Crowhurst, 1991) has led to $n_1 = 46/H$, where H is the height in metres. Thus, a building of height 70 m is predicted to have $n_1 = 0.65$ Hz. Floors are, however, more systematic. A composite slab floor on simple constant-depth beams will have a natural frequency approximately $n_1 = 22\sqrt{gd/L^2}$, where L is the larger spacing of the column grid and d is the depth of the beams on this span, measured from the mid-depth of the slab. Thus a 15 m span with $L/d = 20$ gives $n_1 = 4.0$ Hz. A higher value of the numerical constant applies if the floor is designed for an exceptionally high live load. Further formulations for the estimation of natural frequency are given by Blevins (1979).

It has been pointed out above that the 'real' structure may be significantly different from the design model which takes conservative estimates of continuity and connection stiffness, and ignores the stiffness of claddings, etc. A common-sense appraisal of such factors is desirable; this should include a realistic appraisal of coexistent 'live load' mass rather than the specified value. The stiffness of concrete members should be assessed using a dynamic Young's modulus; $E_c = (20 + 0.3f_c) \text{ kN/mm}^2$ is a general guide, where f_c is the characteristic strength in N/mm^2, or BS8110 may be consulted.

Damping and dampers

Natural damping is often the most difficult property to predict for dynamic analysis. The intrinsic damping of steel or prestressed concrete structural elements is relatively low; bolted steel or ordinary reinforced concrete structures (ignoring the further effects discussed below) are still likely to show only a log dec of 0.03. Higher damping depends on energy dissipation through the foundation, or in non-structural components (cladding, partitions, etc.). The latter will depend largely on frictional effects, and it is a key characteristic of frictional damping that there will be an activation threshold; below the amplitude of oscillation at which the friction forces reach the limiting friction value there is no slip and thus no contribution to damping (Wyatt, 1977).

Full-scale measurements generally are costly and often difficult to obtain at appropriate stress levels and strain rates; there are few free-decay results, and many reported values have been inferred from response to ambient excitation, generally at much lower levels than relevant to a design check. Caution is therefore necessary. Note that there are several alternative measures (cf. see page 7), which are readily confused. For consistency, the suggestions that follow are all expressed as logarithmic decrement, δ. Further background and discussion is given by Warburton (1992) and Blevins (1977, Chapter 8).

Values of damping to be assumed in checking aerodynamic stability have been given as $\delta = 0.03$, 0.04, 0.05 for steel, composite and concrete bridges, respectively (Department of Transport, 1993). For lattice towers, the UK Code, BS8100, gives comprehensive guidance, starting from a base value varying from $\delta = 0.015$ for welded fabrication to $\delta = 0.08$ for fully black-bolted fabrication with cleaned but otherwise untreated joint faces. Allowance is then made for energy dissipation in the foundation by applying a multiplying factor which varies between 1.0 and 3.0, although an additive allowance would seem more rational. In strong winds there is also an aerodynamic damping which is commonly at least as large as the basic structural damping.

Comprehensive guidance for steel chimneys, noting the role of the lining, is given in draft ENV 1991-2-4 (1991). This ENV also gives values for concrete chimneys and for buildings which are proportional to the predicted natural frequency, for example $\delta = 0.045 n_1$ for steel buildings. There is a rationale here, in that taller buildings will have lower natural frequency and also lower relative influence of damping-effective 'non-structural' components. $\delta = 0.075 n_1$ for concrete chimneys is, however, arguably controversial; other authorities have advocated $\delta = 0.06$. The SCI guide (Steel Construction Institute, 1989) suggests $\delta = 0.16$ for office composite floors, but this might be non-conservative for industrial floors supporting machinery.

Damping can be dramatically enhanced by dampers based on

— impact devices;
— viscoelastic materials;
— friction devices;
— tuned mass dampers.

The first three are relatively simple and low maintenance approaches and successful applications have been made to masts and to footbridges (Brown, 1977). Larger devices on similar principles, possibly mobilizing steel plastic deformation in place of friction, have been developed to control earthquake deformation of buildings. The tuned mass damper (TMD) does require highly optimized design and aftercare (Warburton 1992), but offers the possibility of a very high value of δ independent of amplitude. It is wise to allow for some departure in service from the nominal characteristics of the components, but even so an auxiliary mass as little as 2% of the structural mass can ensure the equivalent of $\delta = 0.20$ for harmonic excitation. The TMD is somewhat less effective for random excitation as the spectral response bandwidth is wider than with a simple viscous damper that would give the same equivalent damping of harmonic excitation. The multiple tuned mass damper (MTMD) is a recent development that is much less sensitive to the achievement of optimal parameter values (Abé and Fujino, 1994).

Useful calculations can also be made of foundation damping, as described in Key (1988, Chapter 9). There is considerable current interest in active damping, using forces generated from an external power source under computer control (Warburton, 1992), but such developments are beyond the scope of this guide.

Acceptance criteria: ultimate limit state

A familiar basis for checking the provision of adequate strength is the ultimate limit state (ULS) check incorporated into the so-called limit state design code format. Acceptance is conditional on

$$\gamma_s S_k < R_k \gamma_m \tag{3.5}$$

in which S_k is the characteristic value of load effect, S, R_k is the characteristic value of resistance (strength), R, and γ_s, γ_m are partial factors applied to S and R, respectively.

In principle, a probability level is associated with S_k and R_k, and this is well established for R_k, commonly set as the value that would fail to be achieved in 1 in 20 of an 'ensemble' of structures that were notionally built to the designed strength. It is not so uniformly established for loads; only relatively slowly are Codes defining the explicit probability levels as the basis for S_k. It is still common for the load to be specified on the basis of a return period (average time interval between events exceeding the selected value) comparable to the intended life of the structure, although there are exceptions, of which the most important is earthquake effect (see page 30).

If the dynamic effect causes a moderate increase in response to an important established load-case, such as wind-gust action, the calculated maximum dynamic response stress could be added in to S_k, but there are three complicating factors:

— the code value γ_s is broadly empirical, derived from experience which very probably includes some dynamic response, although no explicit allowance was included in the design equations;

— the ULS check is presumably focused on inelastic response, and although inelastic behaviour enhances energy dissipation, there may be a problem with cumulative deformation;
— peak static and dynamic responses may not be concurrent.

Three different approaches are exemplified in UK Codes making provision for wind-gust dynamics. For buildings, BS6399 Part 2 includes an explicit but very broad-brush dynamic contribution. An *ad hoc* analysis can thus readily be substituted, but there is no provision to compel this for structures within the stated scope of the Code. At the other extreme, for cooling towers, BS4485 Part 4 requires *ad hoc* dynamic analysis but provides a simple formulation for the response output. For lattice towers, BS8100 Part 1, includes a small implicit dynamic contribution together with a simple sensitivity check parameter. If this is not satisfied, a full dynamic analysis is required. Engineering judgement and common-sense are necessary when incorporating estimates of dynamic response into codified procedures of these types.

The design-check problem is fortunately much more straightforward where the dynamic problem is dominant. For earthquake action, well-developed explicit procedures (see page 30) have been codified, permitting integration with static design codes through an equivalent static load, or full *ad hoc* dynamic analysis, in appropriate cases. Potentially catastrophic aerodynamic instability of bridges must be ruled out by reference to an enhanced wind speed focusing on an appropriate probability of excedence (Department of Transport, 1993).

Harmonic excitation in resonance with a natural frequency should generally be avoided, although this is not always possible if the excitation frequency is not fixed (for example, periodic vortex shedding in wind or water currents). For example, a guide for wind-turbines (Stam, 1994) calls for a margin of 20% on frequency, i.e. avoiding the condition $0.8n_1 < n < 1.2n_1$ (where n is the exciting frequency) (zone C in Figure 6). If $n < 0.3n_1$ or $n > 1.4n_1$, no special allowance is required (zones A), and the intermediate cases (zones B) can be treated by application of the appropriate DAF (Figure 6). Resonance would also be likely to cause problems with the criteria addressed in the following section.

Acceptance criteria: other limit states

Metal fatigue

It is obvious that dynamic response leads to cyclic fluctuation of stresses, so there is a *prima facie* cause for concern about fatigue whenever the excitation can be sustained over lengthy periods of time, or can occur sufficiently often to accumulate a large number of stress cycles. The discussion that follows is primarily focused on steel elements; light alloys are commonly more fatigue sensitive: reinforced and prestressed concrete less so. It is difficult to generalize about structural plastics, especially where very high cycle-counts at low stress ranges are involved. A good general discussion of fatigue and other phenomena is given by Collins (1993).

Most steel structures must be deemed to include defects from which fatigue cracking can potentially propagate. The Miner's quotient assessment of cumulative damage is the most common approach. Test data are commonly fitted by the expression $N = c/S^m$, where N is the life (cycle count) at a constant stress range S. The index m is at least 3, and the implied rate of accumulation of damage is thus highly sensitive to the stress range. Under constant-amplitude vibration in a favourable (generally indoors) environment there may be a limiting range S_0, below which any number of cycles can be tolerated; this may permit acceptance of the effects of out-of-balance high-speed machinery giving an extremely high cycle-count at low (non-resonant) stress levels.

Under random excitation giving a very broad spread of values of S, or under most conditions of outdoor exposure, the existence of a limiting range is questionable, and most specifications require consideration of damage by stress cycles below the limiting value S_0 that would apply in more favourable circumstances but permit the use of a higher value of the index m in this range. In all cases the 'true' elastic stress distribution must be presumed, either by applying a stress concentration factor (SCF) to the conventional 'engineers' prediction of stress, or by using a value of the constant c in the SN relationship derived from tests on specimens that can be presumed to include similar stress concentrations. The latter is common for many details of structural fabrication and connection, but an SCF is always required for checking tube-on-tube welded joints (BS5400 Part 10, etc.).

In such cases of response to random excitation, the resonant-frequency 'narrow band' is commonly important, and there is an analytical solution for the total contribution to Miner's quotient per unit time (i.e. summing over the continuous modulation of amplitude in this response; Barltrop and Adams, 1990). An excellent rule-of-thumb is to consider a range $S = 4.5\sigma_N$ (where σ_N is the RMS narrow-band stress) operative at a frequency $\frac{1}{3}n_1\tau$, where τ is the duration of the dynamic response and n_1 is the response (narrow-band) frequency.

For many structures exposed to long-term random dynamic loading (lattice towers, bridges, grandstand roofs, for example) the eventual fatigue check may centre on a nominal stress amplitude as low as $20\,\text{N/mm}^2$ (range $S = 40\,\text{N/mm}^2$, possibly further enhanced by an SCF), with a cycle count as high as 10^7 cycles. It is worth noting that excitation operative for only 1% of total time causing response at 1 Hz will produce such a count in 30 years. In practice, such a combination is rare unless either:

— the ULS design has taken advantage of substantial inelastic modification of the elastic stress distribution, or
— the response is likely to be sustained by specific resonance phenomena, for example, vortex shedding under wind.

In view of the sensitivities noted above, it is desirable to think in terms of a safety margin on S (rather than on predicted life). This margin should be substantial unless it can confidently be assumed that any cracking would be detected at an early stage.

Human reaction and response to vibration

Human perception of vibration is derived from several senses, varying from the sense of touch, through perception of the 'inertia loads' on various organs and the balance organs in the ear, to the muscular efforts developed to retain position. The resulting composite sense is very acute and operative over a very wide range of frequencies. The sensitivity is such that the level at which inertia forces would cause immediate injury is at least 1000 times the threshold of perception (Irwin, 1978). The governing limits are thus commonly a question of the levels causing feelings of insecurity or interruption of concentration, which are greatly variable between persons (Chen and Robertson, 1972) and according to the circumstances of exposure, and are correspondingly difficult to codify. A high-quality residential environment may require limiting the RMS sway of the building for one-hour duration at five-year return period to a value less than 10 times the typical threshold of perception (Lawson, 1982).

At an intermediate level, a limit based on fatigue-decreased proficiency at skilled tasks is easier to establish objectively. For a one-hour exposure, the limit is about 40 times the typical threshold. It is now believed that this limit is strongly affected by the exposure time but it is not clear whether a similar dependence applies at the lower level discussed above. An overall limit of three times the skilled-task limit may be applicable on health and safety grounds.

The 'circumstances of exposure' referred to above include:

— other concurrent distractions (noise, movement of persons or vehicles);
— concurrent activity of the subject (dancing, walking, eating, resting, for example) in decreasing order of magnitude of motion for the same subjective reaction, perhaps a factor of two for each of the steps;
— experience and acclimatization;
— expectation of the subject of environmental quality.

'Baseline' curves of acceleration as a function of frequency are commonly presented as three straight segments on a double-logarithmic plot (Figure 9). At frequencies of the order of 3 Hz, where the dominant sensor is the balance mechanism, the curve is horizontal; i.e. over a range of frequency, the same acceleration is associated with a given reaction level. At frequencies above about 10 Hz, a given cyclic velocity is identified with reaction level, and at very low frequencies the distinctive parameter is probably the rate of change of acceleration. Recommendations for specific applications can be developed by defining the RMS acceleration as a factor R times the base curve. For example, ISO 10137 and BS6472 recommend limiting R for continuous vibration as follows:

— critical working areas (operating theatre, precision laboratory, etc.)   1.0
— residential (daytime, 2–4 permissible) at night   1.4
— offices   4
— workshops   8.

Much higher values are permitted in offices and workshops for intermittent or impulsive vibration, for example, several occurrences per day up to $R = 128$; $R = 20$ is then suggested for night-time residential buildings. The building-sway recommendation has been cited. Floor vibrations are discussed in an SCI design guide (Steel Construction Institute, 1989). Much higher levels are suggested for footbridges (BS5400). Some comparative values and broad subjective descriptions are given in Figure 9.

In all cases, however, the inherent human sensitivity to low levels of motion, and its corollary of relatively slight change in subjective response per unit change in amplitude

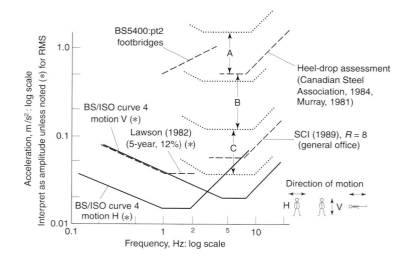

Figure 9 Human subjective response to vibration. Subjective descriptions: A, strongly perceptible, cracking of brittle finishes likely; B, clearly perceptible, admissible for coarse manual labour; C, perceptible, no effect on normal buildings

of motion, must be borne in mind. The SCI recommendation for floors at (say) 6 Hz corresponds to an amplitude of less than 0.1 mm, less than 1% of the likely permitted value of static live-load deflection. If a subjective-reaction problem proves to arise, however, vigorous changes will be necessary; anything less than a reduction by a factor of two is unlikely to give a significant improvement in reaction. Changing to the next larger serial size of structural element will not usually suffice.

Initial design

The foregoing sections have indicated that designers faced with dynamic loading do not start with a clean sheet on which they can impose their will. A recapitulation of the principal distinctions between static and dynamic design is in order:

— in a dynamic response there is not a single response for a given loading – the response varies with time and frequency;
— primary and secondary parts of the structure may interact.

The question of primary and secondary components here refers to the cases where the element at issue is secondary, in the sense of having a much smaller mass than the primary structure by which it is supported. There are numerous practical cases: antennae or slender stacks on a building, flexible or flexibly-mounted equipment or pipework, flarestacks on offshore platforms, and so on. The mass ratio may be 100:1 but if the natural frequencies are comparable, interaction may be significant. A procedure to assist identification of cases where further study is required is given in a guide to design of nuclear structures (American Society for Civil Engineers, 1986).

The following points should also be borne in mind (a further commentary can be found in Hitchings, 1992):

— the maximum displacements may not occur at the same positions as they would for a static analysis;
— effects of non-structural mass must be included;
— the medium surrounding the structure (e.g. water) can contribute to both the dynamic damping and mass;
— for most structural problems damping is low and the precise mechanisms are ill-defined;
— joints/boundaries can be difficult to model in a dynamic analysis (often flexible and/or ill-defined);
— maximum dynamic response is usually greater than the maximum static response for the same order of magnitude of loading.

Most civil engineering structural problems yield a preliminary design without reference to dynamics, the notable exceptions being long-span bridges and deep-water platforms. An estimate can then be made of the dynamic parameters of the structure, which can be assessed against the potential excitations; this is discussed further in Chapter 4.

If there is a harmonic excitation, resonance should be avoided, and it may be both desirable and practicable to change the natural frequency by design change. If not, it may be possible to reduce the excitation or increase the damping to achieve an acceptable result. Acceptance criteria may include fatigue or subjective comfort. The same may apply in cases of frequent, repeated or sustained random excitation. In these cases the output from the dynamic analysis will usually be an RMS value, applied to similar ULS or serviceability checks including fatigue and/or comfort. In most cases, it is favourable to increase the stiffness of the structure. These stages are set out more formally by Bolton (1994, Chapter 10).

Where the critical dynamic event is of short duration, the energy-absorption capability of the structure, or ductility, may be of greater importance than the natural frequency. This reaches its extreme in the case of impact of vehicles or vessels on structures, a problem not taken further in this guide, but designs for blast and seismic effects are addressed specifically in the next chapter.

4. Specific dynamic loadings

Wind-induced vibrations of structures

This section deals with two broadly distinguishable mechanisms of dynamic excitation of structures: the short-term variations of wind speed, i.e. *gusts*, and the excitations that can occur even in smooth flow which are categorized as *aerodynamic instabilities*.

Wind-gust actions

In storm-force winds in temperate climates, the turbulence created by flow past obstacles on the ground generates a characteristic gust structure, the 'neutral atmospheric boundary layer', in which the wind speeds experienced by a structure vary randomly, but (at the crucial peak period of a storm) are statistically stationary for a period of the order of an hour. The power-spectrum (Fourier integral transform) approach is thus ideal for most problems of the resulting dynamic response of civil engineering structures.

The gust structure is to be visualized as carried past the structure at the mean wind speed $\bar{V}$ and gust size and gust duration are thus in direct proportion. In the spectral representation, the concept of gust size is replaced by that of correlation between the spectral components at frequency n Hz at points separated by a distance λ, which is expressed by the 'normalized co-spectrum' $R(\lambda, n)$ (the crisper term 'coherence' is frequently used for $R(\lambda, n)$ although this is strictly incorrect). $R(\lambda, n)$ is a non-dimensional quantity, unity for small separations, with a roughly exponential decay as λ increases. Many authors indeed use an exponential fit as a basis for computation, but a more complex Bessel-function formulation (Harris and Deaves, 1981; Irwin, 1979) has been developed from the theory of homogeneous isotropic turbulence (HIT – the condition in a free field, ignoring the distortion caused by ground proximity). The HIT functional forms are preferable for many applications, but the distinction has little effect on the simple solution for alongwind response considered here.

Large gusts entrain a cascade of progressively smaller components, which give a highly characteristic upper tail to the wind speed spectrum (Figure 10(a)), and it is in this range that structural resonant frequencies regularly occur (moored or tethered buoyant structures excepted). In this range the above visualization applies very well, the normalized co-spectrum is a universal function of $n\lambda/\bar{V}$, and the effective distance L_n over which the gust action at that frequency can be regarded as well correlated is given by $L_n = \bar{V}/cn$ in which c is a non-dimensional factor. For the crosswind dimension of the alongwind component of gust speed the value $c = 8$. This means that the 'resonant gust' is small compared to the loaded length of the structure, h (say). For example, a

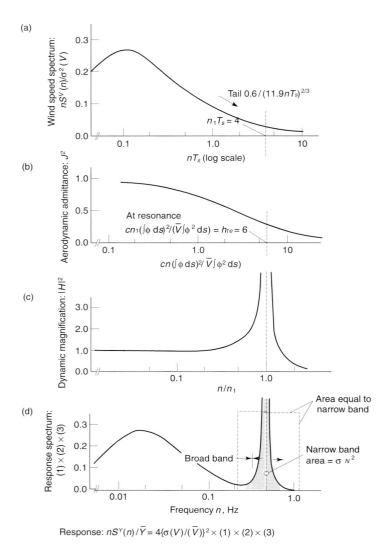

Figure 10 The sequence of spectra and transfer functions for wind-gust response analysis. (Example. Structural parameters: cantilever, $H = 60$ m, natural frequency $n_1 = 0.5$ Hz, shape function $\phi = z/H$ $(\int ds)^2 / \int \phi^2 ds$ $= 0.75H = 45$ m. Wind parameters: $V = 30$ m/s, $T_s = 8$ s, $c = 8$)

tower 60 m high is likely to have natural frequency $n_1 = 1.0$ Hz; in a wind $V = 29$ m/s this gives $L_n = 4$ m. Resonant dynamic response is thus much smaller than would be given by a fully-correlated excitation; the reduction factor is called the 'aerodynamic admittance' J^2 (say). When $L_n \ll h$ there is a simple approximation to J^2, which is presented below and in the accompanying illustrative example. A lower frequency than normal for this size of structure has been used for Figure 10, thus increasing the dynamic sensitivity and making the presentation clearer.

A further effect tending to reduce dynamic response is aerodynamic damping. The effect of the mean wind speed is enhanced when the structure is moving against the wind, and reversed half a cycle later; in the simplest form this dissipates energy equivalent to logarithmic decrement $\delta = P/mn_1\bar{V}$, in which P is the mean wind load and m is the structural mass (or load and mass per unit length, respectively). This can be added to the structural natural damping, and is commonly of significant benefit to lattice (skeletal) structures and to many other structures which would otherwise be sensitive to gust action.

Specific dynamic loadings – wind

The parameters describing the wind-gust excitation are:

— the mean speed, $\bar{V}$;
— the RMS gust speed $\sigma(\bar{V})$ (or the intensity of turbulence, $\sigma(\bar{V})/\bar{V}$);
— the spectrum $S(n)$, describing the distribution of $\sigma^2(\bar{V})$ with frequency;
— the cospectrum $R(\lambda, n)$, replaced for practical purposes by $L_n = \bar{V}/cn$.

In the UK code for buildings, BS6399 Part 2, V is determined from a 'map' speed V_b by application of factors S_c and T_c. The intensity of turbulence is given directly as S_t or $S_t T_t$. The results of the extensive CIRIA field study of wind structure (Harris and Deaves, 1981) fitted in an analytic framework form the basis of ESDU (1991). The approximation given for ordinates in the upper tail of the spectrum (*op. cit.*, equations 6.2 and 6.3) can be further simplified by inclusion of the ESDU function 'A' in a modified timescale τ_s (say), used with the Harris–von Karman upper-tail formulation, i.e.

$$nS(n)/\sigma^2 = 0.6/\tilde{n}^{2/3} \tag{4.1}$$

in which $\tilde{n} = 11.9 n\tau_s$. This estimate is conservative relative to the full expression (*op. cit.*, Appendix B), but the difference is only significant when the resonant frequency is unusually low, such as compliant offshore structures. Representative values of τ_s are given in Table 1. These values are compatible with BS6399 Part 2 and are broadly implicit in the dimensional formulation given in BS8100 Part 4 (guyed masts). In non-uniform terrain, V and $\sigma(V)$ can be evaluated from the tables in BS6399 Part 2 and used in conjunction with τ_s for the rougher terrain. The 'roughness length' z_0 is commonly used in the literature to quantify the terrain roughness.

The analysis sequence is illustrated in Figure 10 (Davenport, 1961, 1962), i.e. the ordinates of the wind-speed spectrum are multiplied successively by the aerodynamic admittance J^2 function and by the mechanical admittance function which expresses the dynamic augmentation. Since the spectrum relates to the mean-square of the respective variable, the latter is the square of the harmonic response function, H^2/K^2 (where K is the modal generalized stiffness). The generalized form of the independent variable is shown in each case; the plots are then aligned on frequency n_1. The format of plotting $nS(n)$ to a logarithmic scale of n retains the visual interpretation of the area as the mean-square (variance, σ^2) value; change of the scaling parameters ($\tau_s, \bar{V}, h, n_1$) simply causes a lateral shift of the universal curves shown.

The response spectrum thus comprises two parts, one being a narrow band of frequencies centred on the natural frequency, the other a broad band of lower frequencies where the response is closely quasi-static. Only the former is considered

Table I Modified timescale τ_s (seconds)

Reference windspeed V_b		25 m/s			32 m/s		
Location (terrain)		Coast	Country	Town	Coast	Country	Town
Roughness length z_0 (m)		0.003	0.03	0.3	0.003	0.03	0.3
Height above ground, z (m)	10	4.4	3.5	2.8	3.6	2.9	2.1
	20	6.3	6.0	5.4	5.6	5.0	4.2
	50	7.3	9.2	10.6	7.0	8.4	8.7
	100	7.9	11.0	15.1	7.3	9.8	12.7
	200	8.5	12.7	19.4	7.9	11.4	16.6

Collapse of Tacoma Narrows suspension bridge in 1940, caused by strong torsional oscillation in a wind of around 40 mph

here. The narrow-band contribution to the response variance $\sigma_N^2(Y)$ can be evaluated to a good approximation by taking the peak value of the spectrum (noting the peak $H^2 = \pi^2/\delta^2$) and multiplying by effective bandwidth $\frac{1}{2}n_1\delta$.

For a line-like structure this can be written as

$$\frac{\sigma^2(Y)}{\overline{Y}^2} = \frac{\pi^2}{2\delta} 4J^2 \frac{n_1 S(n_1)}{\overline{V}^2} \qquad (4.2)$$

In the upper tails of the relevant parts of Figure 10, for influence lines in which ϕ is of the same sign throughout, $n_1 S(n_1) = 0.6\sigma^2(V)/(11.9 n_1 \tau_s)^{2/3}$ (given condition $n_1\tau > 1$) and $J^2(n_1) = (2/h_{te})(1 - 1/h_{te})$ (given condition $h_{te} > 4$), in which

$$h_{te} = \left(\frac{cn_1}{V}\right)\left(\int \phi \, ds\right)^2 \bigg/ \int \phi^2 \, ds \qquad (4.3)$$

The above integrals are taken over the full range of the coordinates which defines location on the structure. When account is taken of variation of the wind and wind-resistance parameters over the height of the structure, the algebra becomes extended, as set out by Wyatt (1981), for example, but the foregoing is often sufficient for a sensitivity check, as presented in Illustrative Example 1 (page 63). The importance of $V/n_1 h$ will be noted; the danger signal is a low natural frequency *relative* to typical values for the given size h. A high value of $\overline{V}$ increases the relative significance of dynamic response through both $S(n_1)$ and $J^2(n_1)$.

1:300 scale dynamic model of Jiangyin Yangtze River Highway Bridge during construction in the BMT Environmental Wind Tunnel, Teddington (courtesy BMT Fluid Mechanics Limited and Kvaerner Cleveland Bridge Limited)

Four levels of this procedure are worthy of comment.

(a) The classical 'line-like structure', for which the structural elements are individually slender so that gusts 'see' individual local elements rather than the shape of the structure as a whole, and location is defined by a single coordinate. The above procedures apply directly.

(b) A lattice of slender members, but member locations requiring two coordinates, as for a lattice tower with face width significantly exceeding L_n. The aerodynamic admittance is now expanded to two separation components, evaluated using the free-stream value $c = 8$.

(c) A crude first approximation for non-slender prismatic structures (buildings) by evaluating aerodynamic admittance as in (b). This is well described as the 'lattice plate approach'. It is rational to consider the excitation on front and rear walls separately, but if so, allowance should be made for the empirical observation that gust dynamic pressures on the front face are more widely correlated than the corresponding free-stream wind speeds. A value $c = 6$ or less is appropriate for the front-face dynamic loading. There is little correlation between the resonant frequency components of the forces on the front and rear faces. This level of application is illustrated by an Illustrative Example (see later).

(d) Determination of aerodynamic admittance by wind-tunnel test. This enables account to be taken of lateral and torsional actions, including the effects of flow separation at the corners of buildings which in practice have a major influence.

It is, in principle, possible to establish dynamic response to gusts directly in the wind tunnel, but it is by no means easy to satisfy all the scale requirements. The advantage is automatic inclusion of any feedback from response to the excitation, which may well arise from the effect of motion on the flow separations. The less ambitious experimental determination of the admittance function is much less constrained and offers flexibility in interpretation of change in the various parameters, and is often preferable.

The potential assessment criteria are the conventional ultimate strength (the 'first passage' criterion, presuming failure to occur if the response once reaches a pre-determined threshold), fatigue life, and occupant subjective comfort. Since gust response has to be sustained for at least an hour in the 'design' storm, and at only marginally lower levels in further storms, evaluation of cumulative deformation is much less relevant as a key to economic design than in resistance to earthquakes. It is usual simply to take the expected peak combination of dynamic response with the near-static effect of large gusts, assessed with a conventional load factor. Denoting the narrow-band variance of load effect f as obtained from the modal displacement as $\sigma_N^2(f)$, and the quasi-static (spectral 'broad band') contribution as $\sigma_B^2(f)$, the total variance is $\sigma^2(f) = \sigma_N^2(f) + \sigma_B^2(f)$. The 'design' value is then $\bar{f} + g\sigma(f)$. The most widely used formulation for g is (Davenport, 1964)

$$g = \{2\ln(\upsilon\tau)\}^{1/2} + 0.58/\{2\ln(\upsilon\tau)\}^{1/2} \qquad (4.4)$$

in which $\upsilon = n_1\sigma_N(f)/\sigma(f)$ and τ is the averaging time for V and f, i.e. usually 3600 s in UK practice. In static-response, value $\sigma_B(f)$ can be found by static stochastic-correlation analysis (Wyatt, 1981), by spectral numerical integration or charts, or inferred from the static code such as BS6399 Part 2, as shown in Illustrative Example 1. The problems of reconciliation of this with the conventional design code have been discussed earlier in Chapter 3. Modes other than the fundamental (possibly including fundamental modes in two principal directions and torsion) are rarely important.

The dynamic effect of gusts on the fatigue cycle count is generally bigger than the quasi-static effect of large gusts. Presuming Miner's rule for cumulative damage, there is a straightforward closed-form solution for damage caused by a narrow-band stochastic response (see page 17), but evaluation is difficult where large-gust effects are also important, due to the severe non-linearity of the fatigue process. Fatigue may prove to be a design constraint, especially where local structural restraints (possibly from notionally non-structural components) that are ignored in the ultimate-strength check significantly increase stresses in the elastic range of response (Wyatt, 1984).

Comfort may be an important design constraint. Unfortunately, decision-taking is subject to all the difficulties mentioned on page 17. The *relative* importance of dynamic response in the strength criteria broadly increases with the design wind speed, but this is greatly exacerbated in the *absolute* values relevant for comparison against the comfort criteria; a location in a severe wind environment thus greatly increases the risk of comfort problems. A tall building will have a higher effective wind speed than a lower building at the same location, but it is certainly possible for medium-height buildings to suffer comfort problems.

Aerodynamic instability

It is helpful to identify two different mechanisms of instability, although in practice there are commonly strong interactions between them. First, most slender prismatic structures are subject to forced excitation due to instability of the flow pattern past the structure, resulting in cyclic changes in the wind force, which can occur even on a stationary structure in smooth flow. The dominant example is vortex shedding, which is considered later in more detail. Secondly, some specific structural shapes are susceptible to self-excited oscillation, where movement of the structure causes changes in the wind force tending to augment the motion (Scruton and Flint, 1964).

Two normalized variables are crucial to consideration of aerodynamic stability.

- The reduced velocity $V_R = V/nD$, in which D is the reference dimension of the structure, commonly selected as the cross-sectional dimension of a prism as 'seen' by the wind. The deck width is commonly preferred as the reference for strong self-excited forces on bridges. Critical values of wind speed can be scaled from wind-tunnel tests or other experience by means of this normalization.
- The Scruton number, $K_S = 2m\delta/\rho D^2$, in which m is the mass per unit length of a prismatic structure, δ the structural damping as log dec and ρ is the density of air. Although δ is a non-dimensional quantity, expressing the energy dissipation capability as a fraction of the energy of oscillation, it must be converted for comparison with the aerodynamic input potential in this way to act as a basis for assessment of damping relative to aerodynamic stability.

The simplest self-excitation mechanism, known as galloping, can be treated by a quasi-static analysis. If the structure moves perpendicular to the wind, the vector resolution of the relative velocity of the wind to the structure causes a change in apparent angle of incidence. An observer on a bridge deck moving downwards would perceive the wind as upwardly inclined, and if this produces a downwards change of force, the bridge will potentially be unstable. For the change to act in this 'negative lift slope' sense is uncommon, but by no means unknown. The plain 2×1 rectangle (shorter side facing the wind) is an example; upward relative incidence tends to suppress separation of flow from the lower face, increasing the velocity close to this face and thus decreasing the pressure by Bernoulli's relationship. If the wind forces are high enough, the energy input per cycle will exceed the dissipation by damping and the oscillation will grow progressively, at least until non-linearity intervenes. The critical speed is given by $V_R = 2K_S/(-dC_z/d\alpha)$, where $dC_z/d\alpha = dC_L/d\alpha + C_D$ is the rate of change of the body-axis force coefficient with apparent wind direction, C_L and C_D being the coefficients for the cross-wind ('lift') and alongwind ('drag') force components, respectively.

Bridge deck oscillations commonly involve torsional deformation of the deck, and there is one relevant excitation mechanism for which there is a direct analytical solution. This is 'classical flutter', in which aerodynamic forces couple together vertical and torsional natural modes of oscillation which are otherwise distinct. The phase-lags from the motion to the resulting force and moment, because the flow pattern cannot adapt instantaneously, are crucial to this phenomenon. The analytical solution (Fung, 1955) is based on the flat-plate aerofoil, but gives good results even for many practical box-girder sections, the Severn bridge being the pioneer example. The characteristic of flutter response is very rapid build-up if the critical speed is exceeded and insensitivity to damping; the counter-measure is high torsional resistance to deter coupling. For less well-streamlined sections, strong self-excitation in torsion alone is possible. The resulting characteristics are intermediate between galloping and flutter. There is no strict torsional analogue for galloping and this mechanism has defied pure analytical treatment; wind-tunnel testing remains essential unless the existing body of test results indicates an adequate safety margin. Small changes in structural cross-section can provoke a large change in self-excitation; although this can be valuably exploited through wind-tunnel testing to achieve the required V_R, the designer must critically review the extent to which reliance is placed on scaling and on the actual ambient wind parameters.

The classic manifestation of simple galloping as described above is the overhead cable with cross-section modified by ice accretion. Rain-wind excitation of bridge stay-cables is a more complex phenomenon which currently causes considerable concern, especially in the current practice of bundled tendons within a plastic sheath. The key feature is cyclic change during the oscillation of the angular location on the cable of the rainwater rivulet running down its undersurface (Verweibe, 1998), but a robust predictive analysis has not yet been achieved. Countermeasures include enhancement of damping, interconnection of the stays in the plane of a stay-fan, and various configured surface finishes which aim to control the rivulet (Virlogeux, 1998).

To sum up the impact of self-excited oscillations:

— consideration is only required for relatively few structures, where either a 'wing' can be visualized (bridge decks, slender cantilever roofs, etc.) or where prisms have very low effective damping as expressed by the Scruton number;
— the potential strength of the mechanisms means almost always that occurrence would be catastrophic; the designer must ensure that the critical speed is beyond practical risk levels;
— wind tunnel testing is commonly necessary and is generally a robust and reliable aid.

A detailed specification is available for medium span bridges (Department of Transport, 1993), and a comprehensive discussion of the background is given by Wyatt and Scruton (1981) and Smith and Wyatt (1981). Further background, especially valuable for reports on UK experience at full-scale, is given by Hay (1992). A specific analytic approach aimed at widening the interpretation of wind-tunnel tests has been developed by Scanlan (Simiu and Scanlan, 1986).

In contrast vortex shedding:

— affects virtually all slender prismatic structures or prismatic elements exposed to the wind;
— the critical wind speed for resonance is given by $V = V_R n_1 D$ in which the 'reduced velocity' V_R is a function of the cross-sectional shape (Blevins, 1977). For a circular section, $V_R = 5$. As the natural frequency n_1 is broadly fixed once the outline design of the structure is established, in many cases an economic design forces the designer to accept a resonance condition;
— response amplitudes are inherently limited. Actual values depend on the cross-sectional shape, on the effective damping (Scruton number, see above), and on a number of factors which may perturb the regularity of shedding such as turbulence in the wind (see further discussion below). If resonance is possible, designers must satisfy themselves that response amplitudes will be tolerable.

The regularity of shedding is linked to the correlation of shedding over the length of the prism. Unfortunately, oscillation of the structure has a remarkably strong influence; the 'lock-on' phenomenon. Amplitudes as small as 2% of the diameter may significantly enhance both regularity and correlation of shedding over the length of the structure. The result is that under controlled resonance conditions, as the Scruton number is reduced progressively, the response initially increases progressively, at a rate between $1/\delta$ (in conditions favourable to periodic, correlated shedding, giving a regular sinusoidal force) and $1/\sqrt{\delta}$ (in less favourable conditions favouring a spectral model of the force process). When the amplitude reaches values such that lock-on is effective relative to the prevailing perturbations, in the event of further reduction of damping the amplitude will increase many times more rapidly. The amplitudes then reached would in many cases be unacceptable, especially in terms of fatigue or subjective response.

Many of the structures at risk are of circular or at least round-cornered cross-section, which adds the complication of Reynolds' number dependence. Regular shedding is greatly reduced in the super-critical Reynolds' number range, but the perturbation of the effective Reynolds' number by turbulence and surface roughness is such that this effect has not been well explored to the benefit of designers. Regularity is re-established in the higher trans-critical range. Given the kinematic viscosity of air $v = 1.4 \times 10^{-5}$ and $V_R = 5$, the Reynolds' number at resonance ($n = n_1$) is about $3.5 \times 10^5 n_1 D^2$. The most favourable range of the product $n_1 D^2$ is probably around $1 \, \mathrm{m^2/s}$ (Wootton, 1969).

Thus, if the wind speed for resonance lies within the site wind speed limit, the designer has a difficult problem, unless the Scruton number is high enough to ensure that lock-on is negligible. To check this he or she may use a sinusoidal force model or spectral methods. The latter are generally more optimistic but should be used with caution if critical wind speeds can credibly occur in conditions of low turbulence, for example in conditions of atmospheric temperature inversion. The threshold windspeed, above which such conditions are unlikely, is unlikely to exceed 15 m/s in the UK climate, lower over rough terrain, but possibly higher where air temperatures are higher than the ground surface temperature.

The analytic model for sinusoidal crosswind force f is $f = \frac{1}{2} \rho V^2 D C_L \sin 2\pi n t$, where C_L is a coefficient of alternating crosswind force ('lift'). At resonance this gives the normalized response amplitude $\eta = y/D$ of an SDOF system as $\eta = C_L V_R^2 / 4\pi K_S$. Extension through modal analysis and conversion to an equivalent static load is straightforward. For the circular prism, the value taken for C_L must take account of the factors introduced above; values around 0.7 sub-critical and 0.3 trans-critical are likely. The UK lattice tower code (BS8100) follows this exactly, with refinements in presentation. The Eurocode (ENV 1991-2-4 Appendix C) is similar, but adds the refinement of applying the excitation only to a limited length of the structure, to allow for imperfect correlation. To allow for lock-on this length is increased from $6D$ to $12D$ as a function of the amplitude of motion. It is perhaps surprising that the threshold for lock-on action is set as high as $y = 0.1D$, but length $6D$ may be envisaged as a conservative estimate.

The stochastic model postulates a power spectrum S_{CL} for the lift coefficient, defined by the variance $\sigma^2(C_L)$ and a bandwidth parameter; the spectrum for crosswind force per unit length is thus $S_f = (\frac{1}{2} \rho V^2 D)^2 S_{CL}$. The most common choice of spectral shape is a Gaussian curve giving the peak ordinate as $n S_{CL}/\sigma^2(C_L) = 1/(B\sqrt{\pi})$, where B is the bandwidth parameter. The analysis proceeds exactly as described for gust action, replacing L_n by an empirically-determined correlation length L_c. Values $\sigma(C_L) = 0.15$, $B = 0.3$, $L_c = 1.0D$ to $1.5D$ have been suggested for conditions of 'normal' turbulence at high Reynolds' number (Vickery and Basu, 1984).

The ESDU data sheets set out a sophisticated spectral analysis with two distinct models (ESDU 1986, parts 1 and 2). The model applicable when the response is small and lock-on is negligible (amplitude less than about 0.015D) is broadly compatible with the proposals of Davenport and Vickery (Vickery and Basu, 1984), but the large amplitude model remains controversial. A distinctive feature of (ESDU, 1986a) is explicit allowance for the effects of turbulence and surface roughness on the interpretation of Reynolds' number.

Concrete chimneys are likely to have predicted response below the main impact of lock-on, and a small-amplitude computation model with a simple allowance for lock-on treated as a negative aerodynamic damping which diminishes the Scruton number provides a good practical procedure. Based on (Vickery and Basu, 1984), the Canadian code (National Research Council, 1985a, b) includes special provisions recognizing the possibility of low turbulence if the critical speed is low. A more complex version is given in American Concrete Institute (1995). Extremely slender steel chimneys may be able to accept the relatively large response locked-on. Most steel chimneys come into the sensitive region, however, and in the UK it is general to fit spiral strakes which effectively destroy the vortex-shedding excitation at the expense of roughly doubling the conventional design wind load (Walshe and Wooton, 1970). Increasing use is being made of auxiliary dampers.

A difficult question is posed by the possible vortex-induced vibration of members in welded tubular towers because in many cases the optimal structural selection will give a critical speed within the local design value, and a resulting Reynolds' number in the low super-critical range. A steel tubular member 9 m long, 273 mm in diameter would have a natural frequency of 21 Hz fixed-ended, perhaps 19 Hz in practice, and thus critical speed 26 m/s, Reynolds' number 5×10^5. Damping values as low as $\delta = 0.01$ have been reported for such elements, which with a wall thickness of 8 mm gives $K_S = 11$. Most of the guidance cited above would suggest a response posing a serious fatigue risk given these values in a severe-climate location, but experience with members in this range has been generally satisfactory. Attention has been drawn above to the probability of a favorable range of Reynolds' number (or of $n_1 D^2$, in this example 1.4); unfortunately the guides are inconclusive or conflicting on this point.

The final point to make about vortex shedding concerns proximity effects. Vortices shed and carried downstream may influence a neighbouring structure more severely than the structure of origin. This is particularly of concern if the structures are identical (pairs of chimneys, for example) as lock-on will then coincide and redouble, but serious problems can also occur when lighter modern construction is placed in the wake of an existing structure. Vortex excitation of cables is routinely controlled by damping, often by the 'Stockbridge' damper, of which there are proprietary variants. Proximity effects on multiple cables are commonly controlled by spacers incorporating elastomer inserts to enhance damping.

Thus, for many problems the designer is likely to seek specialist advice but considerable uncertainty may well remain. For many established examples of problems in service there are apparently similar cases where no problem has been experienced. For special cases (non-circular sections, special locations) the wind tunnel will be called upon, but the above problems of interpretation are likely to remain.

Earthquake-induced vibration of structures

The following is intended as a brief introduction to the earthquake-resistant design and analysis of structures. As with other sections of this guide, the subject is a much wider one than can be covered within the few pages of this section; further guidance is given in general textbooks on the subject (see, for example, Dowrick, 1987; Key, 1988; Naeim, 1989) and by the detailed references cited below.

Quantification of earthquake hazard

The main source of earthquake hazard is violent ground motion, and the response of structures is always essentially dynamic in nature – that is, inertial forces predominate. There are other hazards associated with earthquakes, such as fault rupture, landslip, and tsunami (more commonly known as tidal waves); secondary hazards such as fires may also be triggered by an earthquake. These other hazards do not usually give rise to dynamic responses and will not be discussed further here.

Specific dynamic loadings – earthquake

Structural failure of a building in Kobe, Japan (from EEFIT, 1997)

A qualitative indication of the ground motion hazard for a particular site can be obtained from the Swiss Reinsurance atlas (1978). More precise quantification of design ground motion (for which many current codes specify a return period of about 500 years for normal buildings) can be obtained from the zoning maps in national codes of practice (International Association for Earthquake Engineering, 1996; McGuire, 1993; Paz, 1994). Where these sources may be out of date, or where local features such as active faults or unusual foundation soils render them inappropriate or where special facilities such as nuclear power stations are involved, a site-specific assessment of seismic hazard (Kramer, 1995) is usually necessary. Ground motions close to the causative faults of large earthquakes have special features (Bolt, 1995); these are allowed for approximately by the 'near-source' factors in the Uniform Building Code (UBC, 1997).

The soil overlaying the bedrock at a site may have an important influence on the amplitude and frequency content of ground motions. In particular, soft soils can very significantly amplify motion in the low frequency range. Codes of practice invariably make allowance for this; for example, UBC (1997) provides a simple tabulated method, based on five standard soil profiles. Analytical techniques are also available (Pappin, 1991; Hisada and Yamamoto, 1996; Ejiri and Goto, 1996; Kraemer, 1995). The possibility of soil liquefaction (loss of strength in loose, saturated granular soils under cyclic loading) may need to be considered; it is discussed by Earthquake Engineering Research Institute (1994) and more detailed methods for determining liquefaction potential are given in Eurocode 8 Part 5 (ENV 1998, 1994–8) and an ISSMFE manual (1993).

Earthquake response spectrum

The amplitude and period content of design ground motions is often given by a 'response spectrum'. This represents the value of the peak response of a linear elastic single degree of freedom (SDOF) system with viscous damping when subjected to a specific ground motion record. The structure's natural period (T or $1/n$) is conventionally taken as the abscissa, and curves are drawn for various levels of

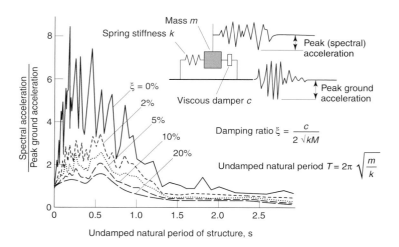

Figure 11 Response spectrum for El Centro 1940 earthquake (from Booth, 1994)

damping (Figure 11). There are similarities to the steady-state response to harmonic ground motion (Figure 6), which would also show no magnification of acceleration at zero period and attenuation at long period. However, earthquake ground motions are highly random in frequency content and are of finite duration, so that earthquake response spectra are significantly different from harmonic response curves, in particular giving much lower amplification of response at resonance, particularly for low damping ratios.

It should be noted that the response spectrum gives no information about the duration of response (hence number of damaging cycles) that the structure experiences, which can have a very significant influence on the damage sustained. Further discussion of response spectra is provided in the section *Analysis for earthquake effects* (page 34).

Design strategies

(a) Conventional solutions. Conventional earthquake resistance design makes use of standard structural forms, with appropriate measures to prevent the risk of life-threatening collapse in a rare, but intense earthquake supplemented by measures to minimize the damage sustained in a more moderate but frequent, event. This strategy accepts that in the rare earthquake irreparable damage may occur, a possibility that is implicit in most seismic code provisions but which is not necessarily understood by building owners, occupiers and insurers or indeed by the general public.

The main strategy for preventing collapse has traditionally been provision of ductility. This is the opposite quality to brittleness, and may be defined as the ability to sustain repeated excursions beyond the elastic limit without fracture. Due to the cyclic, imposed displacement nature of earthquake loading, a ductile structure can absorb very large amounts of energy without collapse; the designer must think in terms of designing for maximum imposed displacements, rather than imposed loads.

Achieving ductility is partly a matter of choosing the right structural system, and partly a matter of detailing. In the former category comes the important concept of 'capacity design', as described by Paulay (1993). This involves ensuring a hierarchy of strengths within a structure to ensure that yielding occurs in ductile modes (such as flexure) rather than brittle modes (such as buckling or, for reinforced concrete, shear). There are other aspects of structural form which are important, particularly regularity in elevation (to avoid 'soft' or weak storeys) and regularity in plan (to minimize torsional response). These aspects are described in many textbooks and are quantified in some codes of practice, including Eurocode 8 (ENV 1998, 1994–8) Part 1.2, section 2.2.

Detailing of the structure is also important to ensure ductility. For concrete structures, this primarily involves reinforcement detailing and in steel structures connection detailing. The latter aspect has been particularly recognized following the failure of welded connections in the Northridge earthquake of 1994 (Burdekin, 1996). The primary reliance is on empirical solutions to these problems, as described in codes of practice, such as Eurocode 8 (ENV 1998, 1994–8) Part 1.3 and UBC (1997). Textbooks discussing these issues for concrete include Paulay and Priestley (1992), Booth (1994) and Penelis and Kappos (1996). Textbooks for steel include Mazzolani and Piluso (1996), but the subject is currently in a state of some flux, following failures of steel structures in recent earthquakes; see Burdekin (1996) and Federal Emergency Managment Agency Guidelines 267 (1995).

Another important aspect of detailing is to allow for the maximum inelastic deflections caused by the design earthquake. Non-seismic resisting elements of a structure such as cladding and infill walls must be able to accommodate these deflections safely, as must (crucially) the gravity load bearing structure, which still suffers the seismic displacements even when not contributing to seismic resistance. In addition, adequate separation between adjacent structures must be provided. Codes of practice (e.g. Eurocode 8 and UBC) give guidance on suitable limits.

Recent earthquakes have proved the success of such strategies in protecting life, but there is increasing awareness that protection of property and economic worth is also important, and insufficiently recognized in some current codes of practice. The enormous economic losses in recent earthquakes confirm this; losses were £15 billion in the 1994 Northridge, California earthquake and at least £70 billion in the 1995 Kobe, Japan earthquake. The next generation of US codes is expected to be based on 'performance-based' principles which address these issues more fully (SEAOC, 1995).

(b) Active and passive control systems. Alternative strategies of designing for earthquake resistance involve modification of the dynamic characteristics of structure to improve seismic response. Their theoretical basis is described by Warburton (1992); Booth (1994, Chapter 10) also provides an introduction to the subject. The systems can be classified as either *passive* or *active*.

The most common type of passive system involves lengthening the structure's fundamental period of vibration by mounting the superstructure on bearings with a low horizontal stiffness; this is known as base or seismic isolation. Where this increases the fundamental period above the predominant periods of earthquake excitation, the acceleration (but not necessarily displacement) response is significantly reduced. Usually, additional damping is provided in the seismic isolation bearing to control deflections.

The principle of seismic isolation is illustrated by Figure 12. The reduction in response, often of the order of 50%, has proved highly effective in recent earthquakes in reducing damage to both building structure and building contents. Skinner *et al.* (1993) and Kelly (1996) provide textbook treatment. UBC (1997) provides codified guidance for seismic isolation of buildings while AASHTO (1991) and Eurocode 8 (ENV 1998, 1994–8) Part 2 treat bridge structures. Seismic isolation has been incorporated in many hundreds of recent structures, particularly in bridges and also in

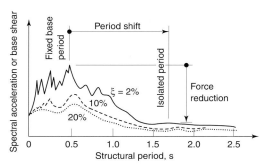
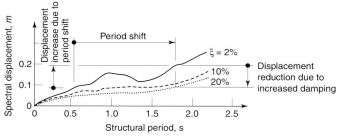

Figure 12 Effect of seismic isolation on forces and displacements for an earthquake with predominant period around 0.5 s (from Booth 1994)

buildings such as hospitals with contents that must remain functional after an earthquake. It has also been used to improve the seismic resistance of existing structures.

Another form of passive system is provision of additional structural damping in the form of discrete viscous, frictional or hysteretic dampers (Earthquake Spectra, 1993).

Active systems modify the dynamic characteristics of a structure in real time during an earthquake, by computer-controlled devices such as active mass dampers. Presently, very few buildings are actually constructed in this way, but there has been a recent large international research effort (Casciati, 1996; Kabori, 1996; Soong, 1996). Due to their adaptability, active systems are less dependent for their effectiveness on the precise nature of the input motion (a concern for passive systems, particularly where they are very close to the earthquake source) but they must have a very high degree of reliability to ensure they function during the crucial few seconds of an earthquake.

Analysis for earthquake effects

Analysis is only one part of the design process; conceptual design, detailing and proper construction are the other vital components for ensuring good seismic performance. This section provides a brief theoretical review of the basis for seismic analysis, describing the main analytical techniques currently used by designers. More complete details are given by Clough and Penzien (1993), a standard general text for dynamic analysis, and Chopra (1995), which deals specifically with concerns for earthquake engineers.

(a) Basic equations of motion for linear single degree of freedom (SDOF) systems. Consider the linear SDOF system shown in Figure 3 subjected to a time varying ground displacement $z(t)$. Let the relative displacement of the system to the ground be $y(t)$; y is then the extension of the spring and dashpot. From the equation of motion, it follows that:

$$m(\ddot{y} + \ddot{z}) = -ky - c\dot{y} \qquad (4.5)$$

Rearranging equation (4.5), and replacing m, k and c by the system's radial frequency ϖ and damping ratio ξ gives:

$$\ddot{y} + 2\xi\varpi\dot{y} + \varpi^2 y = -\ddot{z} \qquad (4.6)$$

Given a description of the input motion $z(t)$ (for example, from an accelerograph recording), the solution of equation (4.6) provides a complete time history of the response of a structure with a given natural period and damping ratio, and can also be used to derive maximum responses for constructing a response spectrum (Figure 11). Due to the random nature of earthquake ground motion, numerical solution techniques are needed for equation (4.6), as described by Clough and Penzien (1993).

(b) Basic equations of motion for linear multiple degree of freedom (MDOF) systems. The dynamic response of many linear MDOF systems can be split into decoupled natural modes of vibration, each mode effectively representing a SDOF system. A modified form of equation (4.6) then applies to each mode, which for mode i becomes:

$$\ddot{Y}_i + 2\xi_i\varpi_i\dot{Y}_i + \varpi_i^2 Y_i = \frac{L_i}{M_i}\ddot{z} \qquad (4.7)$$

Here Y_i is the generalized modal response in the ith mode. (L_i/M_i) is a participation factor, which depends on the mode shape and mass distribution, and describes the participation of the mode in overall response to a particular direction of ground motion. For a two-dimensional (2D) structure with n lumped masses, responding in one horizontal direction

$$\frac{L_i}{M_i} = \frac{\sum_{j=1}^{n} \phi_{ij} m_j}{\sum_{j=1}^{n} \phi_{ij}^2 m_j} \qquad (4.8)$$

In equation (4.8), ϕ_{ij} describes the modal displacement of the jth mass in the ith mode. The higher modes often have very low values of (L_i/M_i), and their contribution can then be omitted. In this way, the computational effort is greatly reduced. In cases where only the first mode in each direction is significant (often the case for low- to medium-rise building structures) equivalent static analysis may be sufficient, as described later.

(c) Response spectrum analysis of linear SDOF systems. With a knowledge of the natural period and damping of a SDOF system, its peak (i.e. spectral) acceleration S_a can be determined directly from an appropriate response spectrum. In undamped systems, this peak response occurs when the equivalent spring is at its maximum extension point, so that the maximum force in the spring is given by

$$F = mS_a \qquad (4.9)$$

From equation (4.9), the peak (i.e. spectral) displacement S_x of the spring is given by

$$S_x = \frac{F}{k} = mS_a \frac{T^2}{4\pi^2 m} = \frac{S_a T^2}{4\pi^2} \qquad (4.10)$$

For structures with relatively small viscous damping, the same relationships are still approximately true, because the maximum acceleration occurs when the velocity is low and hence the damping force (which is velocity proportional) is also low. Therefore, equation (4.9) and hence also (4.10) are still very good approximations for lightly damped systems.

Thus the two most important parameters of structural response – maximum force and displacement – can be determined for a linear SDOF system directly from the acceleration response spectrum, provided only that the mass, natural period and damping are known.

It is important to realize that the spectral acceleration S_a is an absolute value (the true acceleration of the structure in space) whereas the spectral displacement S_x is a relative value, measured in relation to the ground, which itself is moving in the earthquake. This at first sight may seem confusing, until it is remembered that the absolute acceleration of the mass is determined by the force on it (equation (4.9)) which itself is determined by the relative compression of the spring with respect to the ground (equation (4.10)).

(d) Response spectrum analysis of linear MDOF systems. By considering the response of each mode separately, a response spectrum analysis is also possible for an MDOF, if generalized modal quantities are used (compare equations (4.6) and (4.7)). For example, for a 2D structure with n lumped masses, responding in one horizontal direction, equation (4.9) is modified to give the maximum base shear in the ith mode as:

$$F_i = \frac{\left(\sum_{j=1}^{n} \phi_{ij} m_j\right)^2}{\sum_{j=1}^{n} \phi_{ij}^2 m_j} S_a = m_{eff,i} S_a \qquad (4.11)$$

where S_a is the spectral acceleration corresponding to the damping and frequency of mode i.

Higher modes with low effective masses $m_{eff,i}$ may contribute little to response and can usually be neglected. Since the sum of effective masses $m_{eff,i}$ of all modes equals the total mass, a good test of whether the first r modes are sufficient to capture response adequately is:

$$\sum_{i=1}^{r} m_{eff,i} \geq 0.9 \sum_{i=1}^{n} m_i = 0.9 \text{ (total mass)} \qquad (4.12)$$

A response spectrum analysis gives the maximum response of the structure for each mode of vibration considered. Although it is rigorously correct to add the response in each mode at any time to obtain the total response, the maximum responses in each mode, calculated from response spectrum analysis, do not occur simultaneously, and hence simple addition produces an overestimate of response. A common and usually adequate approximation is the square root of the sum of the squares (SRSS) rule, where the maximum total response is estimated as the SRSS combination of the individual modal responses. However, this may be unconservative for closely spaced or high frequency modes, and other methods, such as the complete quadratic combination (CQC) method, are available (see Gupta, 1990).

There are many commercially available computer programs which can perform response spectrum analysis, and it is now regarded as a standard rather than a specialist technique.

(e) Ductility modified response spectrum analysis. For yielding SDOF systems, ductility modified acceleration response spectra can be drawn, representing the maximum acceleration response of a system as a function of its initial (elastic) period T, damping ratio ξ and displacement ductility ratio μ (μ is the ratio of maximum displacement to yield displacement). The reduction in acceleration response of the yielding system compared with the elastic one is period dependent; for structural periods greater than the predominant earthquake periods, the reduction is approximately $1/\mu$, for very stiff systems there is no reduction, while at intermediate periods a reduction factor between $1/\mu$ and 1 applies.

To derive peak accelerations and internal forces, the system can be treated as linear elastic and the ductility modified spectrum used exactly like a normal elastic spectrum. However, deflections derived from this treatment must be multiplied by μ to allow for the plastic deformation.

It is now standard practice to analyse MDOF systems in the same way. That is, a yielding MDOF system is treated as elastic, and an appropriate ductility modified spectrum is substituted for an elastic one. Acceleration and force responses are derived directly and deflections are multiplied by μ. However, this procedure is (contrary to the case for SDOF systems) not rigorously correct; although it gives satisfactory answers for regular structures, it can be seriously in error for structures (such as those with weak storeys) where the plasticity demand is not evenly distributed. Nevertheless, most codes of practice allow the use of ductility modified spectra for design, and give appropriate values for the reduction factors (called q or behaviour factors in Eurocode 8 and R factors in UBC) to apply to elastic response spectra. Displacement-based design (see page 39) overcomes some of the drawbacks of using ductility modified spectra.

(f) Smoothed design spectra. Due to the highly random nature of earthquake ground motions, the response spectrum for a real earthquake record contains many sharp peaks and troughs, especially for low levels of damping. The peaks and troughs are determined by a number of uncertain factors, such as the precise location of the earthquake source, which are unlikely to be known precisely in advance. Therefore, spectra for design purposes are usually smoothed envelopes of spectra for a range of different earthquakes; indeed, one of the advantages of response spectrum analysis over time history analysis is that it can represent the envelope response to a number of different possible earthquake sources from a single analysis, and is not dependent on the precise characteristic of one particular ground motion record. Codes of practice such as UBC (1997) and Eurocode 8 (ENV 1998, 1994–8) provide smoothed spectra for design purposes.

(g) Non-linear time history analysis. The most rigorous form of dynamic analysis involves stepping a non-linear model of the structure through a complete time history

of earthquake ground motions. The advantage of the method is that it can give direct information on non-linear response, on duration of response (and hence number of loading cycles) and on the relative phasing of response between various parts. One disadvantage of the technique is that it can be complex and difficult to interpret, involving much more computing power than the simple methods. Also, the analysis is only as accurate as the algorithms it contains for non-linear material response, which may well be subject to uncertainty. In general, time history analysis remains a specialist task.

The analysis must be performed for a number of different earthquake time histories to reduce dependence on the random characteristics of a particular record; Eurocode 8 requires the use of at least five records. Extensive databanks of real ground motions are commercially available; alternatively, artificial accelerograms may be generated from design spectra, although the phasing and energy content of such records may be unrealistic for some purposes.

(h) Soil structure interaction (SSI). Structural analyses usually assume that ground motions are applied via a rigid base, thus neglecting the effect of ground compliance on response. (The effect of soft soils on the ground motions themselves, however, is invariably allowed for – see page 31). Although this rigid base assumption may lead to an underestimate of deflections, it is usually conservative as far as forces are concerned, because ground compliance reduces stiffness and usually moves structural periods farther from resonance with the ground motion. However, this conservatism may not always apply, and Eurocode 8 Part 5 lists the following cases where SSI should be investigated:

(a) structures where $P-\delta$ (second order) effects need to be considered;
(b) structures with massive or deep-seated foundations, such as bridge piers, offshore caissons and silos;
(c) slender tall structures such as towers and chimneys;
(d) structures supported on very soft soils with an average shear wave velocity less than 100 m/s.

Allowance for SSI effects is usually a specialist task; further information is given by Pappin (1991) and Wolf (1985, 1994).

(i) Equivalent static analysis. This is the type of analysis presented in most contemporary codes of practice and is conditional for its accuracy upon response being dominated by one mode of vibration in each direction. The maximum lateral base shear is first calculated; equation (4.13) gives the relevant formulae in UBC (1997). Other current codes follow similar formats.

$$V = \frac{C_v I W}{RT} \quad \text{but } V \leq \frac{2.5 C_a I}{R} W \text{ and } \geq 0.11 C_a I W$$
$$\text{In addition, } V \geq \frac{0.8 Z N_v I}{R} W \text{ (high seismicity Zone 4, only)}$$
(4.13)

where

V = ultimate seismic base shear (force units, e.g. kN)
C_v, C_a = seismic coefficients, depending on the zone factor Z (see below) and the soil profile = $0.8Z$ to $3.2Z$ in UBC
I = importance factor = 1 to 1.25 in UBC

R = reduction coefficient depending on the ductility of structure = 2.8 to 8.5 in UBC
T = first mode period of the building (seconds)
W = building weight (force units, e.g. kN)
Z = zone factor expressed as the peak ground acceleration on rock (in gravity units) for a 475-year return period = 0.075 to 0.4 in UBC
N_v = factor allowing for proximity to active faults = 1.0 to 2.0 in UBC.

(V/W) represents the shape of a standard design response spectrum with a peak amplification on ground acceleration for 5% damping of 2.5 and a minimum value at long period to allow for the uncertainty in long period motions and for proximity to active faults.

The base shear calculated by these methods is then applied to the structure as a set of horizontal forces, with a vertical distribution based on the first mode shape of regular vertical cantilever structures. Horizontal distribution follows the mass distribution, with some additional allowance for torsional effects. A worked example of analysis to UBC (1997) is given on pages 68–71.

(j) Choice of analysis method. Equivalent static methods are usually adequate for conventional, regular building structures under about 75 m in height. A response spectrum analysis is required for taller buildings, because higher mode effects may become important, and also for buildings with plan or elevational eccentricities, because torsional effects or non-standard modeshapes may be significant. Codes of practice such as Eurocode 8, and UBC (1997) specify the degree of eccentricity at which such analysis is required.

Unusual or very important structures may require non-linear time history analysis, and this may also be required where the inaccuracies implicit in the use of ductility modified response spectrum analysis become unacceptable.

'Displacement-based' design combines some of the realism of non-linear time history analysis, while avoiding most of its complexities. It is current recommended US practice in the seismic strenthening of existing buildings (Federal Management Research Agency, 1997) and is likely in future to form the basis for code design of new structures (Fajfar and Krawinkler, 1998). An extensive discussion of displacement-based design is provided in ATC 40 (Applied Technology Council, 1996).

Codes of practice

Seismic loading codes for 41 countries are reproduced in the World List (International Association for Earthquake Engineering, 1996) and discussion of many codes is given by Paz (1994). However, the material-specific rules for earthquake-resistant design (covering for example detailing and dimensioning) are equally important, and are not included in either of these; direct reference to the codes must be made for these data.

Comparison with wind loading

There are fundamental differences between wind and earthquake loading which have important implications for design. The most important differences are summarized in Table 2.

Dynamics: an introduction for civil and structural engineers

Table 2 Main differences between wind and earthquake loading

Characteristic	Wind	Earthquake
1. Source of loading	External force due to wind pressures.	Applied base motion from ground vibration.
2. Type and duration of loading	Wind storm of several hours duration. Loads fluctuate, but are predominantly in one direction.	Transient cyclic loads of at most a few minutes total duration. Loads change repeatedly in direction.
3. Predictability	Good statistical basis is generally available.	Poor.
4. Sensitivity of loading to return period	Moderate; +15% typical for ×10 on return period.	High; maximum credible earthquake often greatly in excess of 'design' values.
5. Influence of local soil conditions	Little effect on dynamic sensitivity.	Soil conditions can be very important.
6. Spectral peak input range	Gust: <0.1 Hz.	Usually 1–5 Hz.
7. Main factors affecting building response	External shape of building or structure. Generally only global dynamic properties are important. Dynamic considerations affect only a small fraction of building structures.	Response is governed by global dynamic properties (fundamental period, damping and mass) but plan and vertical regularity of structure also important. All structures are affected dynamically.
8. Normal design basis	Elastic response is required.	Inelastic response is usually permitted, but ductility must be provided.
9. Design of non-structural elements	Applied loading is concentrated on external cladding.	Entire building contents is shaken and must be appropriately designed

Vibration induced by people

Introduction

All movement by people results in fluctuating reaction forces between the structure and the person. Even simple walking causes a modest cyclic change in the height of the body mass above the floor, and the product of the mass with the respective accelerations equates directly to a cyclic force. This effect is clearly much larger in many forms of rhythmic activity; the movements can be much larger, and it is likely that many people will be moving in synchronism. The typical walking-pace frequency of somewhat less than 2 Hz is also a central frequency for most rhythmic activities, and generally it is essential to ensure that there are no structural modes of vibration that have a shape receptive to these forces that also have a natural frequency close to this value. However, the excitation will certainly include Fourier harmonics, and it is commonly necessary to consider the resonant magnification of one or more such components. It may be necessary to consider horizontal as well as vertical excitation.

Since the effective stressing of the structure arises from the inertial effects of resonance, and human tolerance of accelerations within the conventional built environment are relatively low, the serviceability criterion of subjective response has prime impact on acceptability. This may be only a question of complaint resulting in loss of rental values for office floors, but could potentially result in panic in crowded dance arenas or similar places of assembly. Lateral (horizontal) stability may also be at issue. Strength checks have, however, recently been specified in the UK for places of assembly where rhythmic activities are possible.

Rhythmic activities

The frequency and magnitude of excitations resulting from dancing, jumping or aerobics are essentially limited by the requirement of keeping to the beat. Clearly the minimum net reaction between a person and the floor is zero, so vigorous activity results in a cycle that includes a free-flight phase. During the contact phase, the force exerted on the floor commonly follows a fairly smooth curve, similar in appearance to half a sinusoidal cycle. It proves that the ratio of the duration of the contact phase (t_p,

Specific dynamic loadings – people

Rhythmic excitation, including harmonics, may cause floor resonance

say) to the periodic time (T_p), i.e. the contact ratio $\alpha = t_p/T_p$, is a significant parameter for description of the vigour of the activity. The maximum value of the contact force is $\pi G/2\alpha$, in which $G = mg$, where m is the mass of the body, irrespective of the periodicity of the activity. A Fourier representation for the complete time history of contact force $F(t)$ is then

$$F(t) = G\left[1 + \sum r_N \sin\left(2\pi Nt/T_p + \phi_N\right)\right] \quad (4.14)$$

in which

$$r_N = \{2(1 + \cos 2\pi N\alpha)\}^{1/2}/(4\alpha^2 N^2 - 1) \quad (4.15)$$

Numerical values of r_N and ϕ_N are given in Table 3 (Ji and Ellis, 1994; BS6399 Part 1).

In order to define the loads fully, the frequency range of the activity must be defined, as well as the density of the crowd. Individuals can sustain motion up to 3.5 Hz, but for

Table 3 Fourier coefficients for various activities involving jumping

Activity	Contact ratio α		Fourier frequency factor N			
			1	2	3	4
Pedestrian movement,	2/3	r_N	1.29	0.16	0.13	0.04
low-impact aerobics		ϕ_N	$-\pi/6$	$\pi/6$	$-\pi/2$	$-\pi/6$
Rhythmic exercises,	1/2	r_N	1.57	0.67	0.00	0.13
high-impact aerobics		ϕ_N	0	$-\pi/2$	0	$-\pi/2$
Normal jumping	1/3	r_N	1.80	1.29	0.67	0.164
		ϕ_N	$\pi/6$	$-\pi/6$	$-\pi/2$	$\pi/6$
High jumping	1/4	r_N	1.89	1.57	1.13	0.67
		ϕ_N	$\pi/4$	0	$-\pi/4$	$-\pi/2$

larger groups, co-ordinated movement at the higher frequencies is difficult and the frequency range ($1/T_p$) of 1.5 to 2.8 Hz is appropriate. The references cited above also suggest a reduction factor of 2/3 can be applied to the excitation defined in Table 3 to allow for lack of synchronization between individuals.

The detailed solution to obtain the maximum response for a particular given T_p is algebraically tedious. In practice, where a subjective comfort criterion is at issue, the potential resonant magnification when a Fourier component coincides with the structural frequency is so large that this can be presumed to be the governing case. The case of $\alpha = \frac{1}{2}$ which gives $r_3 = 0$ is addressed in Illustrative Example 3. For consistency of notation, the structural natural frequency will be denoted by its period, T_0 say, and $N = T_p/T_0$ takes such integer values as arise within the specified limits on activity frequency. The steady-state response amplitude generated by the resonant Fourier component is then

$$\tilde{y} = \frac{G}{k}\frac{\pi r_N}{\delta} \qquad (4.16)$$

in which k and δ is the structural stiffness and the damping logarithmic decrement, respectively. The reduction factor for lack of crowd synchronism is not included in this value. Modal analysis can be used as appropriate. Greater caution is required when stress criteria are at issue, to include non-resonant effects occurring during each contract phase (Figure 7). Fourier components at least up to $N = 3$ should be considered.

Vertical jumping also generates a horizontal load which may be critical for some structures, such as temporary grandstands. A horizontal load of 10% of the vertical load has been suggested, although less onerous limits have been suggested for the natural frequency of modes susceptible to horizontal excitation than for vertical modes.

Serviceability of floors subject to normal pedestrian use

It was shown in the previous section that in normal use the second and third Fourier components of the contact force may be significant. Walking (say) eight paces applies 24 cycles of the third component and can thus result in resonant build-up of response not far short of the steady state if the floor natural frequency is less than about 7 Hz. Modern economical floors are likely to have a natural frequency less than 7 Hz if the span exceeds about 10 m, whether of steel or concrete construction. Such floors commonly behave as highly-orthotropic plates; in conjunction with ill-defined boundary conditions, the analysis is complex, but the resulting prediction of the subjective response parameter (R, say) is broadly inversely proportional to the floor mass per unit area (m_f, say). A design guide prepared in the UK (Steel Construction Institute, 1989) suggests that if the natural frequency is between 4.8 and 7.0 Hz, the value

$$R = \frac{85\,000}{m_f A \delta} \qquad (4.17)$$

should be limited to 8 for general purpose offices. In this formulation, A is the effective participating area of the floor, defined for a number of different floor layouts in the guide cited (the product $m_f A$ must be expressed in kilograms). Guidance is also given on appropriate values of the damping logarithmic decrement, δ.

For floors of natural frequency between 3.0 and 4.8 Hz, potentially subject to second-harmonic resonance, greater caution is suggested. The relatively low value of the second harmonic factor (r_2) in Table 3 for $\alpha = 2/3$ is not necessarily conservative for walking where there is no 'free flight', especially for vigorous walking at a high-pace frequency such as would affect floors of frequency around 4 Hz.

For floors of natural frequency exceeding 7 Hz, the most significant feature of the excitation is the small impulsive action at each footfall. This causes a relatively large immediate local response which decays rapidly as energy is dispersed over the floor as a whole. A similar inverse relationship, $R \propto 1/m_f$, is predicted. The 'heel drop' impulsive–excitation test has been extensively advocated in North America as a technique for the assessment of such floors – see Canadian Standards Association (1984) and Murray (1981). It should be noted that the traditional timber floor is very responsive to the footfall impulsive action; its acceptability shows that subjective response criteria at these levels are 'soft'. They are psychological, not directly physiological, and strongly influenced by expectation and experience.

Footbridges

Conventional beam footbridges of spans greater than 25 m are likely to have a fundamental natural frequency within the range of walking-pace frequency. The key feature of this excitation is that the point of application advances by a pace length for each cycle of excitation; the modal analysis must thus take account of the varying value of the mode shape function by which the actual point force is converted to the modal generalized force. As the damping will be very low (for example, $\delta = 0.04$ for composite construction), response will not closely approach the steady state within the number of paces required to cross the bridge. The maximum response can readily be determined by trial, but a valuable closed-form approximation is given in BS5400 Part 2. The latter is based on a force amplitude of 180 N irrespective of frequency, although the fundamental component of walking is now known to increase systematically from about 140 N at 1.6 Hz to 370 N at 2.2 Hz (Blake, 1975, page 19.8).

The result of resonance will generally be found unacceptable, given the low base-level of damping. Friction dampers have been found very effective in practice (Brown, 1977). Most walkers on a sensitive bridge instinctively adjust their pace to the response in a way that maximizes excitation, and the change of natural frequency on reaching the threshold amplitude of a friction device probably inhibits this effect. More commonly, the chosen solution is to change the structural form, usually introducing cable stays, to raise the bridge frequency above the basic pace frequency. There have been no reports of adverse response of a footbridge to the second Fourier component, but there have been reports of adverse transverse response of a stayed bridge carrying exceptionally heavy pedestrian traffic (excitation at one cycle per two paces, left, right, left and so on).

Blast effects

Explosive blast loading

Loading resulting from explosions in general comprises two distinct physical processes: a blast wave-front behind which the air pressure and density differ substantially from the initial ambient values, and the airflow velocities implicit in these density changes. The free-field propagation equations, including the relation between the maximum value of the 'overpressure' (p_s, say), and the concurrent maximum value of the 'dynamic pressure' (kinematic pressure, $\frac{1}{2}\rho V^2$) (q_s, say) were established by Rayleigh and Hugoniot more than a century ago, but the effect of proximity to the ground surface and the interactions on encounter of an obstacle greatly complicates the resulting structural loading. The majority of existing design guidance is based on studies of the effects of nuclear weapons (Newmark and Rosenblueth, 1996); in the events which are more often of current concern, the explosion energy is very much smaller but consideration is focused on correspondingly closer proximity. The result is a significant shortening of the free-field pressure pulse in comparison to the time and spatial scales of the typical structure. This should be borne in mind in the interpretation of much existing guidance.

A universal normalized description of these effects can be given by scaling distance relative to $(E/p_0)^{1/3}$ and pressure relative to p_0, where E is the energy release (kJ) and p_0 the ambient pressure (typically $100\,\text{kN/m}^2$). For convenience, however, it is general practice to express the basic explosive input as an equivalent mass of TNT (W, say). Results are then given as a function of the dimensional distance parameter $Z = x/W^{1/3}$, where x is the actual effective distance from the explosion. W is generally expressed in kilograms (kilotons for nuclear weapons).

Arrival of the blast wave-front is marked by a sharp rise in pressure, followed by a rapid decay. Values of the peak overpressure in free-field blast propagation are given in Figure 13. The effects of an explosion near the ground are enhanced by reflections, corresponding approximately to increasing W by a factor of 1.8. The impulse of the overpressure $i_s = \int p_s(t)\,dt$ is well approximated for $Z > 1$ by $i_s = 0.2W^{1/3}/Z$ (kN s/m^2 for m and kg TNT input units: notation i_s distinguishes this impulse per unit area from the point-load or modal generalized impulse denoted I in Chapter 2).

There is additionally displacement of air to satisfy continuity at the enhanced pressure, expressed through a 'dynamic pressure' $q(t)$, i.e, the kinematic pressure $q = \tfrac{1}{2}\rho V^2$. The maximum value, q_s say, is given by

$$q_s = 5p_s^2/2(p_s + 7p_0) \tag{4.18}$$

The numerical coefficients derive from the basic analytic relationships with substitution $\gamma = 1.4$ for the ratio of the specific heats of air. A full discussion and extensive charts are given by Mays and Smith, 1995.

If the blast wave encounters an obstacle perpendicular to the direction of propagation, reflection increases the overpressure to a maximum value

$$p_r = 2p_s\left\{\frac{1 + 4p_s/7p_0}{1 + p_s/7p_0}\right\} \tag{4.19}$$

The reflection effect dissipates as the perturbation propagates to the edges of the obstacle at a velocity related to the speed of sound (U_s, say) in the compressed and heated air behind the wave front. Denoting the maximum distance from an edge as S

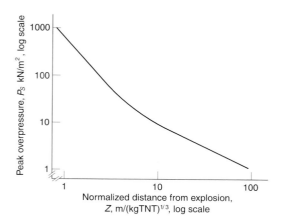

Figure 13 Overpressure as a function of distance from explosion

(for example, the lesser of the height or half the width of a conventional building), the additional pressure due to reflection is considered to reduce from $p_r - p_s$ to zero in time $3S/U_s$. Conservatively, U_s can be taken as the normal speed of sound, about 340 m/s, and the additional impulse to the structure evaluated on the assumption of a linear decay.

After the blast wave has passed the rear corner of a prismatic obstacle, the pressure similarly propagates on to the rear face; linear build-up over duration $5S/U_s$ has been suggested. For skeletal structures the effective duration of the net overpressure load is thus small, and the drag loading based on the dynamic pressure is then likely to be dominant. Conventional wind-loading pressure coefficients may be used, with the conservative assumption of instantaneous build-up when the wave passes the plane of the relevant face of the building, the loads on the front and rear faces being numerically cumulative for the overall load effect on the structure. Various formulations have been put forward for the rate of decay of the dynamic pressure loading; a parabolic decay (i.e. corresponding to a linear decay of equivalent wind velocity) over a time equal to the total duration of positive overpressure is a practical approximation.

Some representative numerical values of blast loads are given in Table 4.

The duration of the positive overpressure phase can be inferred approximately from the value of the impulse; for example, the duration of an equivalent triangular load pulse would be $t_i = 2i_s/p_s$. In the case of a building 50 m from the detonation of 1000 kg of TNT, this would be about 0.03 s. This is clearly small compared to the half-cycle duration of the fundamental mode of anything but the smallest building, and is also much less than the reflected pressure dissipation time for any substantial intact building, for example, if $S = 17$ m, $t_r = 3S/U_s$ is about 0.15 s. The structural implications are discussed later.

Gas explosion loading

The conventional explosives considered hitherto comprise a compact organic fuel pre-mixed with or comprising an oxidant, and on detonation the reaction is virtually instantaneous by comparison with the duration of the resulting load pulse or the natural periods of structural response. The specific energy of combustion of a hydrocarbon fuel is very high (46 000 kJ/kg for propane, compared to 4520 kJ/kg for TNT) but widely differing effects are possible according to the conditions at ignition.

In the circumstances of progressive build-up of fuel in a low-turbulence environment, typical of domestic gas explosions, flame propagation on ignition is slow and the resulting pressure pulse is correspondingly extended.

Table 4 Representative values of blast loads

Explosive charge: TNT equivalent	W	8		1000			kg
Distance from explosion	x	10	50	10	50	250	m
Peak overpressure	p_s	28	2.8	1000	28	2.8	kN/m²
Peak dynamic pressure	q_s	2.7	0.0	1600	2.7	0.0	kN/m²
Peak reflected pressure	p_r	62	5.7	5500	62	5.7	kN/m²
Impulse of overpressure	i_s	0.08	0.016	1.8	0.4	0.08	kN s/m²

Venting following the failure of windows (at typically 7 kN/m^2) generally greatly reduces the peak values of internal pressures. Study of this problem at the Building Research Establishment (Ellis and Crowhurst, 1991) showed that an explosion fuelled by a 200 ml aerosol canister in a typical domestic room produced a peak pressure of 9 kN/m^2 with a pulse duration somewhat over 0.1 s. This is long by comparison with the natural frequency of wall panels in conventional building construction and a quasi-static design pressure is commonly advocated. Strain rate dependence (and the effect of axial loading) as discussed in the next section may, however, be important. Much higher pressures with a shorter timescale are generated in turbulent conditions. Suitable conditions arise in buildings in multi-room explosions on passage of the blast through doorways, but can also be created by obstacles closer to the release of the gas. They may be presumed to occur on release of gas by failure of industrial pressure vessels or pipelines.

Dynamic response to blast

The principal impact of dynamic effects in design of structures to resist blast loading arises where large non-linear (generally inelastic) deformation is accepted. Notionally exact analysis of response is then only possible by step-by-step numerical solution requiring high-quality dynamic finite-element software. However, the degree of uncertainty in both the determination of the loading and the interpretation of acceptability of the resulting deformation is such that solution of a postulated equivalent ideal elastoplastic SDOF system (Biggs, 1964) is commonly used. Interpretation is based on the required ductility factor $\mu = y_m/y_E$ (Figure 14), analogous to earthquake resistance assessment.

The equations for the SDOF system constituting a mode of an elastic system were given in Chapter 2. For example, a uniform simply supported beam has first mode shape $\phi(x) = \sin \pi x/L$ and the equivalent mass is $M = \frac{1}{2}mL$. The equivalent force corresponding to a uniformly distributed load of intensity p is $P = (2/\pi)pL$. The response of the ideal bilinear elastoplastic system can be evaluated in closed form for the triangular load pulse comprising rapid rise and linear decay, with maximum value P_m and duration t_d. The result for the maximum displacement is cumbersome and is generally presented in chart form (US Army, 1969), as a family of curves for selected values of P_u/P_m showing the required ductility μ as a function of t_d/T, in which P_u is the limiting value of the force and $T = 1/n_1$ is the natural period of the structure.

If the pulse duration is short compared to the natural period, the loading can be treated as a pure impulse. The initial velocity $I/M\varpi$ gives kinetic energy $U = \frac{1}{2}I^2/M\varpi^2$. The maximum response is given by equating the area under the load–deformation curve to U. Noting that the modal stiffness is $K = M\varpi^2$, inserting the elastic limit deflection $y_E = P_u/K$ leads to

$$\mu = \frac{y_m}{y_E} = \frac{1}{2}\left(\left(\frac{2\pi I}{P_u T}\right)^2 + 1\right)$$

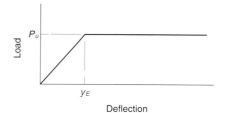

Figure 14 Ideal elastoplastic SDOF system

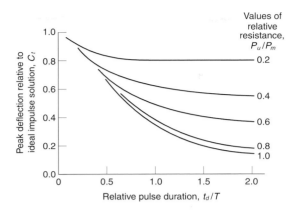

Figure 15 Response of elastoplastic SDF system to linear decay load pulse

For a finite duration ideal triangular pulse $I = 1/2 P_m t_d$, and the chart solutions can be re-expressed by introducing a correction factor C_t so that

$$\mu = \frac{1}{2} C_t \left(\left(\pi \frac{P_m t_d}{P_u T} \right)^2 + 1 \right) \qquad (4.20)$$

C_t is shown in Figure 15.

For most cases of both clad and skeletal structures subject to terrorist or industrial-accident blast it is possible to make a reasonable approximation by evaluating the net impulse and estimation of an equivalent duration of linear decay. The sensitivity of the result to the value of the impulse is noteworthy.

For the commonly important cases where the load duration is less than 0.2 s and substantial inelastic deformation is permitted, the enhancement of the yield stress at high strain rates greatly reduces the deformation. Factors of 1.2 or 1.3 on the static yield stress have been suggested on an empirical basis. As increased yield stress increases the potential importance of local buckling, caution must be exercised in application to elements in compression. A more realistic analysis must take account of the variation of strain rates within the plastic region, both with spanwise location and distance from the neutral axis (Izzudin and Smith, 1996). A logarithmic strain-rate model (Malvern, 1952) can be used for the dynamic yield stress, f_d

$$f_d = f_y (1 + s \ln [1 + \dot{\varepsilon}_p / \dot{\varepsilon}_*])$$

in which f_y is the static yield stress, $\dot{\varepsilon}_p$ is the plastic strain rate, and s and $\dot{\varepsilon}_*$ are material properties; $s = 0.027$ and $\dot{\varepsilon}_* = 10^{-7} \, \text{s}^{-1}$ have been suggested for steel. For a simply-supported beam the effective dynamic resistance P_d becomes

$$P_d = P_u (1 + s \ln [1 + \dot{y}/\dot{y}_*])$$

in which $\dot{y}$ is the rate of deflection and $\dot{y}_* \approx 0.6 s^{1/2} L^2 \dot{\varepsilon}_* / d$ (more refined estimates as a function of load type and section profile are given by Izzudin and Smith, 1996).

To demonstrate this effect, consider a member of span 5 m, depth 350 mm and yield stress 275 N/mm²; this gives $\dot{y}_E \approx 0.02$ m and $\dot{y}_* \approx 4 \times 10^{-6}$ m/s. A natural frequency of 20 Hz would be reasonable, giving $T = 0.05$ s. For a triangular pulse of peak value $P_m = 1.67 P_u$ and duration $t_d = T$, Figure 15 gives $C_t = 0.44$ and thus $\mu = 7$. The average rate of deflection is $\dot{y} = 7 \times 0.02/0.05 = 2.8$ m/s. At this rate, $P_d = 1.36 P_u$, so the effective nominal overstress factor is reduced from 1.67 to 1.23. Repetition of the evaluation of the ductility requirement converges at $\mu = 3$, less than half the value predicted on the basis of the static yield stress.

Machinery

Fundamentals

It is useful to distinguish three classes of machinery:

— rotating, for example turbo-generators, electrical machines and turbines;
— reciprocating, for example most engines, compressors and crushers;
— impact, for example forge hammers and stamping machines.

Machine speeds are usually quoted in revolutions per minute (rpm). Reciprocating machines commonly operate at speeds of 600 rpm or more, rotating machines commonly at higher speeds, with turbo-alternators often running at 3000 rpm to give a 50 Hz supply. Thus, cyclic forces caused by out-of-balance masses rarely have frequencies outside the range of 10–50 Hz and are accordingly much higher than the fundamental natural frequencies of most structural forms considered in this guide. In addition to out-of-balance mass forces at multiples (orders) of the basic operating frequency, reciprocating engines may also exhibit gas forces at half-order increments. These gas forces depend on the engine firing order and the cylinder configuration.

The harmonic forces arising from out-of-balance masses are generally of the form

$$f(t) = Me\varpi^2 \sin \omega t \qquad (4.23)$$

in which M is the rotating mass, e is the eccentricity and ϖ is the machine operating speed expressed in rad/s (radians per second). A related quantity is the balance quality grade G (ISO, 1940-1) equal to the product $e\varpi$. A typical value for turbo-alternator sets is $G = 2.5$ mm/s, equivalent to $e = 0.008$ mm for a 3000 rpm machine ($\varpi = 314$ rad/s). An influential survey is given by Barkan (1962), which remains a primary reference, although it should be recognized that machinery speeds have tended to increase and balance standards have improved considerably over the years since publication. As an example, Barkan suggests $e = 0.05$ mm for a 3000 rpm machine. Forces are generally of the form of rotational and linear loadings (moments and forces) and are usually supplied by the engine builder.

The striking rate of an impacting machine is not restricted to the same degree. The dynamic forces are clearly dominated by the corresponding basic frequency component but a range of higher frequency components reflecting deformation modes of the machine and its mounting are also likely to be significant. Isolation mountings for the machine may well be essential.

Isolation-mounting of machinery

The essential feature of isolation is soft mounting. For a spring-borne mass subjected to a harmonic force of amplitude $\tilde{f}$ arising from an out-of-balance machine running at constant speed n Hz, defining a mounting of stiffness k giving a natural frequency $n_1 \ll n$, the displacement amplitude will be $\tilde{y} = H\tilde{f}/k$. In this frequency range the frequency response function H is little affected by damping and well approximated by $1/(n^2/n_1^2 - 1)$ (cf. equation (2.5)). The force transmitted through the mounting, amplitude $\tilde{f}_T$ say, is thus

$$\tilde{f}_T = k\tilde{y} = \tilde{f}/(n^2/n_1^2 - 1) \qquad (4.24)$$

If n is high, a large reduction is feasible. For example, a mounting with static deflection 5 mm will give $\varpi_1^2 = 4\pi^2 n_1^2 = g/\Delta_w = 196$, which with a 3000 rpm machine ($n = 50$ Hz, $\varpi^2 = 10\,000$ s^{-2}) gives a reduction by a factor of 50. It will be noted

Specific dynamic loadings – machinery

Rotating machinery often yields cyclic forces from 10 to 50 Hz

that this result is independent of the magnitude of the spring-borne mass. However, the displacement amplitude of the machine will be excessive, leading to possible fatigue problems or indeed to transmission of forces through pipework 'bridging' the mounting, if the mass is insufficient.

Reduction in the transmitted force by soft-mounting of impacting machinery depends primarily on achieving a low value of t_p/T_1. The net impulse of the impact (I) is the dominant input parameter; this may be modified by the nature of the workpiece in hammers and press-tools. The displacement response again leads to the estimate of the transmitted force, in this case (noting $k = 4\pi^2 n_1^2 m$) $f_T = 2\pi n_1 I$.

It should be borne in mind that rotating or reciprocating machinery which give modest harmonic excitation in normal operation may also give large forces under fault or emergency conditions. Some rotary machines, such as coal or clinker mills, may give rise to random forces best described by power spectra (compare with wind-gust excitation discussed earlier).

Further guidance can be obtained from Barkan (1962) and the information services of specialist suppliers of machine-mounting components.

Ground-based foundations

The structural design of supports for major items of machinery must normally be checked to avoid resonance (see Chapter 3 and Figure 6). In view of the difficulty of achieving frequencies in excess of 50 Hz for large structures, 'undertuned' foundations, where the fundamental frequency is below the normal machine speed, are now commonly accepted. It is then necessary to ensure that the analysis gives sufficiently accurate modelling of higher modes to confirm freedom from any resonance at the operating speed of the machine, including reasonable allowances for inaccuracy of prediction of input parameters (material properties, connection stiffness, effect of components normally ignored in strength checks, etc.) It is also clear that on start-up and shut-down the machine speed will have to pass through the fundamental frequency.

For ground-based foundations, the elastic compliance of the soil has a major effect on the system dynamic properties. A number of analyses are available giving frequency-dependent relationships between the displacement and force amplitudes in steady-state harmonic motions of a rigid foundation-face under the broad heading of 'elastic half-space solutions'. These are commonly couched in complex-number notation, for example the force is postulated as $f = \tilde{f}e^{i\omega t}$ and the resulting displacement is written $y = (\tilde{f}/k)(a + ib)e^{i\omega t}$, in which k is the respective static stiffness and functions a and b give the solution in non-dimensional ('normalized') form. The latter are usually expressed as functions of the normalized frequency $\Omega = \varpi R/V_s$ in which R is the reference dimension of the foundation and V_s is the velocity of propagation of the shear wave in the soil.

The complex-number format is interpretable in real terms by noting $e^{i\varpi t} = \cos \varpi t + i \sin \varpi t$, and considering only the real part of any resulting products. It will be seen that a expresses displacement in phase with the force, akin to a conventional stiffness, whereas b expresses a component in quadrature, i.e. 90° out of phase. This component results in the force f doing a net work over each cycle, akin to damping. By comparison with the simple dynamic system of Chapter 2, the equivalent damping of a free vibration, expressed as a critical damping ratio ξ is

$$\xi = b/2(a^2 + b^2)^{1/2} \qquad (4.25)$$

This mechanism is radiation of energy from the system by stress waves, and is known as dispersion damping. Energy loss by stress–strain hysteresis in the soil local to the foundation equates to dissipation damping, additive to the dispersion effect.

For high frequencies, Ω of the order of unity, dispersion is of major importance. For example, $n = 8\,\text{Hz}$, $V_s = 200\,\text{m/s}$ and $R = 8\,\text{m}$ gives $\Omega = 2$. For horizontal translation motion of a foundation at the surface of a uniform half-space this gives $a = 0.4$ and $b = 0.5$, whence $\xi = 0.2$, equivalent to $\delta = 1.3$. The same approach is applicable to the soil–structure interaction of a wide range of structures on pad foundations including chimneys, towers and 'gravity' offshore platforms, but in such cases $\Omega < 1$, and dispersion damping is relatively much smaller, especially for rocking motion. Soil hysteresis damping may be more significant.

Solutions are available for horizontal and vertical translation motions (Arnold et al., 1955), for surface and for embedded foundations (Ulrich and Kulmeyer, 1972) and considering the 'off-diagonal' coupling effects between translation and rotation (Veletsos and Wei, 1971) (although these are not normally of major importance). Most solutions relate to circular foundations; equivalent radii are suggested for rectangular foundations (Whitman, 1966). The stiffness solution (function a) is generally regarded as robust, but greater caution is necessary concerning dispersion damping (function b) because non-uniformity of the elastic half-space, especially stratification, leads to partial reflection of stress waves which may seriously reduce the net dispersion. Appropriate solutions are also available for the 'Gibson soil' with progressive increase of stiffness with depth corresponding to a normally-consolidated clay.

Some indication of soil properties suitable for an initial or provisional assessment is given in Table 5. An extended review, giving the full algebraic formulations referred to above, is given by Key (1988, Chapter 9).

Table 5 Dynamic soil properties for initial or provisional elastic analysis

Soil type	Characteristic value	Shear velocity V_s (m/s)	Poisson's ratio, v	Density ρ (kg/m³)	Dynamic stiffness E_d (N/mm²)
	SPT (blows)				
Sand: loose	0–10	100–150	0.15–0.20	1750	50–100
medium	10–30	150–200	0.20–0.25	1800	100–200
dense	>30	200–250	0.25–0.30	1850	200–300
Sandy gravel		100–250	0.15–0.30	2000	50–300
	c_U (kN/mm²)				
Silty clay		100–200	0.20–0.40	1800	50–200
Clay: soft	<20–40	100–200	0.20–0.40	1800	50–125
firm	40–75	125–200	0.20–0.40	1950	100–150
stiff	75–150+	150–200	0.20–0.40	2100	125–200

Note: dynamic Poisson's ratio in water-saturated soils to be treated with caution.

Ground transmitted vibration

Elastic waves

Vibration energy is transmitted through compressible fluids such as air, so-called incompressible fluids such as water and the ground which may range in consistency from loose soils to dense homogenous rock. Although soils may behave in a markedly non-linear manner under very strong shaking, the theory of elastic waves is the starting point in the explanation of how energy travels from a source to a position where it can lead to the imposition of dynamic loads on structures.

The classical wave equation is fundamental

$$\frac{\partial^2 y}{\partial t^2} = c^2 \nabla^2 y \qquad (4.26)$$

where c is constant and $\nabla^2 = \partial^2/\partial x^2$.

This equation and its solutions are described in many texts (e.g. Bullen and Bolt, 1985). There are a number of elastic waves which satisfy the wave equation. One group called body waves are either dilatational or rotational. In dilatational waves (usually called P or compression waves) the disturbance is transmitted through the assumed homogenous isotropic ground at a speed

$$V_p^2 = E(1-v)/\rho(1-2v)(1+v) \qquad (4.27)$$

for an infinite medium, where E is Young's modulus, ρ is density and v is Poisson's ratio and the rotational disturbance (S or shear wave) is transmitted at the speed

$$V_s^2 = G/\rho \qquad (4.28)$$

Shear waves are of two kinds: those in which the particle motion representing the disturbance is in the direction of the propagation of the energy and those in which it is normal to the propagation of the energy. Earthquake strong motion is usually modelled as a shear wave propagating with horizontal particle motion, except very close to an earthquake source.

Another category of elastic waves which exist are surface waves, the most familiar being the Rayleigh wave with a velocity of

$$V_R^2 \approx G/\rho \qquad (4.29)$$

Such waves have their motion concentrated near the surface and are dispersive; that is, their velocity is dependent on wave length and period and a disturbance made of many frequencies changes its shape as it moves away from its source. Since man-made ground vibration affecting the built environment is of this nature the spread of the wave train from the source and the decrease of motion with depth from the surface may affect the way a ground vibration travels from the foundations to a structure itself. Figure 16 shows the ground motion associated with a Rayleigh wave diagramatically.

The geometrical spreading differs for the different waves: body waves travelling in a solid spread and attenuate according to an inverse square law except where they emerge, when they attenuate as the inverse of distance from the source. Surface waves attenuate according to the inverse of the square root of the distance from the source. In stratified ground the picture is complicated because at interfaces waves convert from one to another and are reflected and refracted. Due to these conversion processes, and the lower geometric attenuation, Rayleigh waves are usually of more concern at greater distances from a source. However, close to a source, a driven pile for example, the surface waves may not be distinguishable and attenuation in engineering practice is usually described by empirical equations, for example

$$\Re = k(W^{1/2}/r)^n \qquad (4.30)$$

where $\Re$ is a maximum particle velocity in the wave train (see below) and W is a source energy or energy-related value and r is a distance from the source to a point on the ground surface. Some expressions use a 'slant distance', others radial projection on the ground surface (Ramshaw et al., 1998).

In addition to attenuation by geometric spreading there are scattering and energy dissipating processes which the seismologist calls Q or the quality factor and which can be related to the structural engineer's 'damping' in simple systems as follows:

$$Q = \frac{\pi}{\delta} = \frac{1}{2}\xi \qquad (4.31)$$

Q may vary from less than 10 in soils to over 100 in sound rocks.

Other waves exist at the junction between layers of different elastic properties and in boreholes but it is beyond the scope of this guide to describe them.

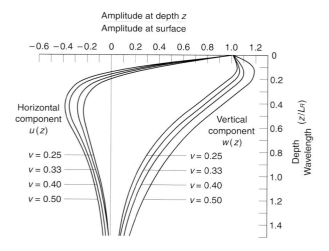

Figure 16 Rayleigh wave amplitudes and depth

Specific dynamic loadings – vibration

Vibrations can be transmitted through the ground, structure and also air

Characterization of ground motion

In the engineering community the strong motion of earthquakes has, for 50 years, been characterized by the 'pseudo-response spectrum'. Other methods have been put forward including power spectra. It is recognized that response spectra do not fully describe ground motion, but in characterizing the less energetic ground motion which is generated by man-made processes for purposes of assessing its impact on structures current codes still retain an even simpler concept, namely, the single kinematic parameter, the most usual being the PPV or peak particle velocity. PPV is either derived as an instantaneous peak vector from three orthogonal measurements at the surface of the ground or as a conservative SRSS (square root of sum of squares) of peaks in the wave train. Current standards which have some bearing on ground dynamics are listed by Skipp (1998). In the UK, guidance on the measurement and evaluation of shock and vibration is offered in three British Standards: BS7385 Parts 1 and 2 and BS5228. In Figure 17 the guidance offered is indicated. BS7385 Part 1 sets out the formal classification of man-made vibrations and indicates the character and levels of ground vibration which might be experienced from a variety of sources.

Many problems arising from man-made vibration, however, relate to human response. Guidance on this is offered in BS6472:1992. Human response in buildings is affected

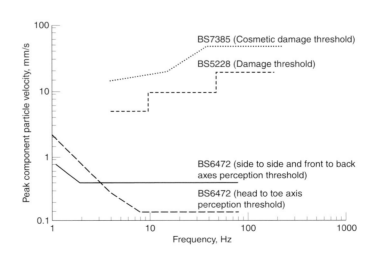

Figure 17 Threshold values for vibration from British Standards

both by structural motion and air motion, i.e. acoustics. This field of 'vibro-acoustics' which seeks to cover the interaction between structure-borne response to ground motion and its coupling to enclosed air is being addressed now for such problems as vibration and noise nuisance from surface and underground railways by ISO technical committee TC108, Sub-Committee 2.

Impact

Introduction

Impact effects on structures arise over a very broad range of circumstances, from high-velocity small-mass missiles to high-mass ship or vehicle collisions with bridge piers. The requirement may be for the structure to withstand the impact without serious damage, or major inelastic deformation may be permitted, including the possibility of sacrificial energy absorbing systems. This introduction focuses primarily on large impact loads producing plastic strains which greatly outweigh the elastic effects.

Impact causes elastic and plastic stress waves, and propagation through the structural thickness can cause failure by spalling. Such effects usually occur within microseconds of the impact, and may be referred to as the 'early time response'. The overall dynamic response of the structure usually occurs on a timescale several orders of magnitude longer, and can thus reasonably be decoupled from the early time response, subject to initial check against spalling. The following treatment addresses single impact events; the effects of repeated impacts are examined in Jones (1989a) and Shen and Jones (1992).

Elementary aspects of inelastic impact

The striking energy of a mass $\mathcal{M}$ impacting with velocity V_0 is $U_0 = \frac{1}{2}\mathcal{M}V_0^2$. If the impact velocity is governed by falling from height h, then $V_0 = (2gh)^{1/2}$, but much higher velocities are possible with debris from an explosion. In some cases, it may be necessary further to allow for loss of gravitational potential energy as the structure deforms, but consideration of absorption of the energy U_0 by inelastic deformation, subject to relatively small elastic rebound energy, covers a wide range of practical examples. There may also be residual kinetic energy, for example in ship collisions; $\mathcal{M}$ in that case must include the virtual water mass (Minorsky, 1959).

Consider a stationary body of mass $\mathcal{M}_1$ struck by a mass $\mathcal{M}_2$ travelling with initial velocity V_2. Assuming that the coefficient of restitution is zero, the masses travel on with common velocity V_3 given by the momentum equation

$$\mathcal{M}_2 V_2 = (\mathcal{M}_1 + \mathcal{M}_2) V_3 \qquad (4.32)$$

A balance of energy for the impact event shows that the loss of kinetic energy, K_ℓ, is given by

$$K_\ell = \frac{1}{2}\mathcal{M}_2 V_2^2 / (1 + \mathcal{M}_2/\mathcal{M}_1) \qquad (4.33)$$

Energy K_ℓ must be absorbed by the structural members and any energy-absorbing system which may be interposed between the impacting masses (Jones, 1989a; Johnson and Mamalis, 1978; Johnson and Reid, 1986). In the extreme case of a relatively large striking mass ($\mathcal{M}_2 \gg \mathcal{M}_1$), the loss of energy during the impact is small.

If the struck mass is constrained to remain stationary throughout the impact event, as a bridge pier, corresponding to $\mathcal{M}_2 \ll \mathcal{M}_1$, the whole initial energy must be dissipated. If an energy absorbing device which has a mean dynamic crushing force P_m is interposed, then the deceleration of the impacting mass is $-P_m/\mathcal{M}_2$. The impacting mass comes to rest after time $T = V_2 \mathcal{M}_2/P_m$, and the crushing distance is $\Delta = \mathcal{M}_2 V_2^2/2P_m$.

Impact behaviour of beams

Given the foregoing presumption that the plastic deformation is dominant, equating the kinetic energy of the impacting mass to the work done in plastic hinge deformation is a simple procedure, following the classical model for static plastic analysis of beams. Thus, a simply supported span of length $2L$, struck at mid-span, in developing a central deflection y_p will create a plastic hinge rotation $2y_p/L$ and the dissipation is $2M_p y_p/L$. Under impact energy $\frac{1}{2}\mathcal{M}_2 V_0^2$, the central deflection is

$$y_p = \mathcal{M}_2 V_0^2 L/4M_p \quad (4.34)$$

If the beam is fully restrained at each end, there are two additional plastic hinges, each with rotation y_p/L, and thus

$$y_p = \mathcal{M}_2 V_0^2 L/8M_p \quad (4.35)$$

If the supports also provide full axial restraint, axial forces will be developed as the transverse displacements increase, and the permanent deformation will be reduced. For a beam of solid rectangular cross-section of depth d (Jones, 1989b)

$$y_p = 0.5d\{(1 + 4\Lambda)^{1/2} - 1\} \quad (4.36)$$

in which $\Lambda = \mathcal{M}_2 V_0^2 L/8M_p d$ is the ratio of the estimate of deflection if bending action only is considered (equation (4.35)) to the depth of the beam. The importance of this effect is shown in Figure 18.

Thus far the analysis has been quasi-static, in the sense that transient effects associated with travelling plastic hinges have been ignored and the static displacement profile assumed to apply throughout. The transverse velocity profile in the transient stage, with travelling plastic hinges at $z = \pm \zeta$ and a stationary hinge at mid-span, is shown in Figure 19(b), compared to the velocity profile in the later stages (which is the same as the static profile) in Figure 19(c).

The quasi-static assumption is used extensively throughout civil and structural engineering and is shown in Jones (1995) to be valid for sufficiently large values of the mass ratios $\mathcal{M}_2/2mL$, where m is the beam mass per unit length, and $2L$ is the total

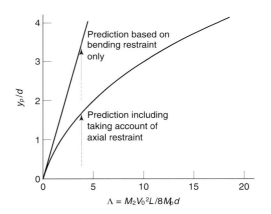

Figure 18 Maximum permanent transverse displacement of a fixed ended beam

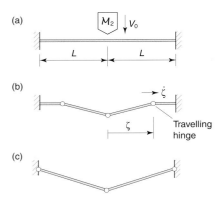

Figure 19 Phases of hinge development in a fixed ended beam

length of the beam. The percentage error (e_w) between an exact rigid perfectly plastic solution and a quasi-static approximation is shown in Figure 20. The error is less than 1% for mass ratios larger than about 33 for the infinitesimal deflection case. If the influence of finite-displacements is retained (i.e. axial membrane effects), then the error is 1% for mass ratios larger than 16.5. Errors of order 10% are obtained for mass ratios as small as 3.25.

The rigid–perfectly plastic approach in the previous section is also valid only provided elastic effects do not play an important role during the beam response. This effect can be assessed by calculating an energy ratio, E_R, which is defined as the external energy divided by the maximum possible strain energy that can be absorbed in a wholly elastic manner by a structure. For a member of length L and cross-section area A

$$E_R = \{\mathcal{M}_2 V_0^2/2\}/\{p_y^2 AL/2E\} \tag{4.37}$$

where E is Young's modulus and p_y is the static yield stress of the beam material. This is a conservative definition of the energy ratio, so it is assumed usually that $E_R > 3$, approximately, is acceptable for the validity of a rigid–plastic analysis.

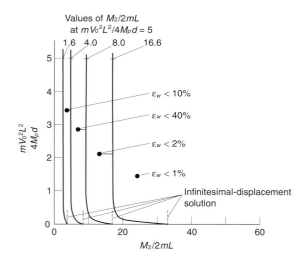

Figure 20 Error intrinsic in quasi-static analysis

Discussion

This section contains a very brief introduction to the behaviour of ductile structures subjected to impact loads producing large inelastic strains. Further information on the accuracy of these methods and their limitations can be found in Jones (1989a). These methods have also been used to study the impact behaviour of a wide range of structures, including beams, frames, plates (Nurick and Martin, 1989), pipelines and shells. The design of energy absorbing systems has been examined by Johnson and Mamalis (1978), for example, while fairly comprehensive summaries of applications in naval architecture and ocean engineering have been given by Jones (1976, 1997). These methods of analysis have been of considerable assistance in the design of various safety features in the automobile industry including impact attenuation devices for motorways (Veillette and Carney, 1988; Carney, 1993).

In some cases the influence of finite displacements are important, as for example, in axially restrained beams suffering maximum transverse displacements larger than the beam depth, as noted in the previous section. In other cases, transverse shear effects can be important, particularly for some of the beams with deep sections used in civil and structural engineering (Jones, 1985, 1989a). In all cases, the dynamic material properties should be used in the various equations rather than the static values, even for quasi-static analyses. A great deal has been written on the strain rate sensitive behaviour of materials (e.g. Jones, 1989a; Al-Hassani and Reid, 1992), but far less emphasis has been given to the dynamic rupture strain (Jones, 1989b, c). Some studies have explored the failure criteria which control the dynamic inelastic failure of structures (Yu and Jones, 1997). A special case of impact failure is that due to perforation and methods of analysis have been developed for small missiles travelling at high velocities and for large masses travelling at relatively low velocities (Corbett *et al.*, 1996; Jones and Kim, 1997; Goldsmith, 1999).

If the impact loadings are repeated, then the displacements in rigid, perfectly plastic structures can grow in certain circumstances, but eventually might reach a limit known as a pseudo-shakedown state (Jones, 1989a; Shen and Jones, 1992). This behaviour can develop due to wave impact loading of ships, marine vehicles and offshore platform structures.

It was remarked above that stress wave propagation effects can be ignored for many structural impact problems. However, it is becoming increasingly clear that wave propagation effects are important during the early phase of motion for predicting the response of certain problems, such as rods and shells impacted axially (Karagiozova and Jones, 1996). Equivalent mass–spring approximate methods can also be used to study the impact behaviour of structures, but these are not discussed here.

There is a considerable amount of activity in this general field at the moment because of the quest to improve safety and produce optimum structural designs and the requirement for accurate hazard assessments throughout industry.

Hydrodynamic loading: wave- and current-induced vibrations

The principles of dynamic excitation of structures by water are closely parallel to those of wind action and the section on wind dynamics should be read as an introduction. Attention is focused here on wave excitation of structures in deep water and on current action on members forming piers or supporting columns for structures in shallow water.

Wave action

The action of waves on deep-water structures such as oil exploitation platforms is generally based on complex non-linear wave models for a design wave with a period of

An oil production platform; dynamic checks include wave loading on primary structure, fatigue action of wind on flare tower and safety of internal blast protection (BP Amoco plc)

order 15 s. Every effort is made to avoid creating structural natural modes with periods close to this value, i.e. natural frequencies in the range of 0.025–0.25 Hz.

Structures where the natural frequencies are above the excluded range are 'fixed', but a design wave of 15 s period will produce a significantly greater maximum load effect on a structure of (say) 4 s period (0.25 Hz) than the simple static value. Allowance can be based on the simple sinusoidal frequency response function, or a better solution taking account of the non-sinusoidal non-linear wave model can be obtained by step-by-step (time domain, Chapter 2) solution. Very large computer models are sometimes used for this purpose.

Structures where the most important natural frequencies are below the excluded range are 'compliant' with low values of restoring stiffness. Since the wave forces depend heavily on the relative velocities the structure tends to follow the horizontal component of the wave motion at the depth where the dominant elements are situated, for example the buoyancy elements of a tension-leg platform. The tension-leg structure is an example which is compliant for horizontal and torsional (yaw) motions but the vertical motion (heave) frequency is likely to lie in the upper range. As well as wave non-linearity, consideration may also have to be given to essentially non-linear

Specific dynamic loadings – hydrodynamic

structural behaviour derived from action of buoyancy forces analogous to an inverted pendulum. This is a specialized field.

Wave fields are generally observed to be significantly random, especially in the 'cascade' of energy down the upper tail of the spectrum. The Pierson–Moskowitz spectrum in a developed form which is fully defined by the 'significant wave height' (or by the RMS surface displacement) is a universal description of the fully-developed sea-state, i.e. where fetch and storm duration are sufficient for the wave field to develop to equilibrium with the driving wind. More detailed forms have been developed for specific locations of limited fetch but with similar upper tails, for example the JONSWOP family for the North Sea.

The response of a fixed structure with natural frequencies in the upper tail of the wave spectrum will therefore show a significant resonant (strictly, spectral narrow-band) component. Analysis is closely similar to wind-gust analysis but with the distinctions:

— wave action is fully correlated down a vertical line; correlation in the along-wave direction follows directly from the wave model and allowance may or may not be made for imperfect correlation in the perpendicular direction (short-crested wave analysis);
— the hydrodynamic forces are modelled by Morrison's equation comprising both drag (proportional to shadow area times velocity squared) and inertia forces (proportional to immersed volume times acceleration);
— hydrodynamic damping is commonly important, but derives from the primary near-static drag forces only.

Inertia forces are clearly relatively more important for a structure of a small number of large elements than for a comparable structure with a larger number of individually smaller elements. For any given Fourier integral component at circular frequency $\varpi = 2\pi n$, velocity amplitude is ϖ times displacement, but acceleration takes ϖ^2. The inertia action is therefore relatively more persistent than drag to higher frequencies. The dynamic response of a 'fixed' structure, with natural frequency in the upper tail of the wave spectrum, is thus more likely to be significant in the 'small number of large elements' case.

Current action

Slender prismatic members in water are subject to forces from vortex shedding. Although many features are common to the discussion earlier in this chapter, the difference by a factor of about 800 in the fluid density means that the Scruton number expressing the effectiveness of the structural damping in this context is likely to be very much smaller than in wind problems. It is therefore generally essential to ensure that resonant conditions cannot occur, by ensuring that the critical speed for the lowest mode is greater than the maximum practical current speed. The special case of flexible 'marine risers' in deep water is beyond the scope of this guide.

Although the principal manifestation of vortex shedding is the cyclic change of force in the crosswind direction, there is also a weaker along-wind fluctuation with a cycle for each vortex shed, i.e. at twice the crosswind frequency, giving $V_R = 2.5$ for circular sections. Resonance at this frequency has proved a problem for piles used as support members for estuarial jetties.

5. Summary of selected design issues

The following is a repeat of selected text which has appeared earlier in this guide. The purpose of this summary is to provide designers with an *aide-mémoire* to revisiting parts of the guide which they may have noted as particularly useful. The extracts below should not be taken out of context nor considered in isolation.

From Chapter 1 – Introduction

(a) The designer can decide whether:
— dynamics is of marginal importance and is safely covered by load factors in conventional checks, or
— dynamics is significant and specific practical design-office procedures should be applied, or
— dynamics is crucial and specialist advice should be considered.
(b) Every effort should be made to avoid resonance.
(c) Where resonance cannot be avoided it may be possible to reduce the magnitude of excitation.
(d) Energy dissipation or damping plays a key role.
(e) There are very diverse potential acceptance criteria for dynamic response.
(f) Human response to the perception of motion is an exceptionally difficult question. It is important that the designer appreciates the problem and can present a clear assessment.

From Chapter 2 – basic dynamics theory

(a) Dynamic loading can be classified into a number of types.
(b) A frequency domain plot can often highlight the most significant characteristics of the loading.
(c) Many civil engineering structures may be treated as linear systems.
(d) The principal cases for non-linear analysis are seismic loading and 'accidental' impact loadings.
(e) A common idealization is to assume that damping is viscous.
(f) Modal analysis is an extremely powerful tool.
(g) There are three principal (response) cases:
— steady-state response to harmonic (sinusoidal) excitation, including spectral (Fourier integral transform) random loading;
— response to transient impulsive or step-change forces;
— complex deterministic force time-histories.

Summary of selected design issues

From Chapter 3: design for dynamic loading – general

(a) Simpler approaches are usually practicable for the lowest mode (natural frequency).
(b) The designer has little power to alter the natural frequency, given an economic design to a conventional structural form. The lowest natural frequency generally follows a defined relationship to the size of the structure.
(c) Disparity of stiffness between the designer's model and reality may be considerable.
(d) A realistic appraisal of coexistent live load mass (is desirable);
(e) stiffness of concrete should be assessed using a dynamic Young's modulus.
(f) Natural damping is often the most difficult property to predict for dynamic analysis.
(g) Full-scale measurements are generally costly and often difficult to obtain appropriate stress levels and strain rates.
(h) Damping can be dramatically enhanced by dampers.
(i) Engineering judgements and common-sense are required when incorporating estimates of dynamic response into codified procedures.
(j) A guide for wind turbines calls for a margin on frequency (to avoid the condition of resonance).
(k) Dynamic response leads to cyclic fluctuation of stresses, so there is a *prima facie* cause for concern about fatigue whenever the excitation can be sustained over lengthy periods of time.
(l) It is desirable to think in terms of a safety margin on S (the stress range) (rather than on predicted life).
(m) The inherent human sensitivity to low levels of motion must be borne in mind.
(n) If a subjective-reaction problem proves to arise, however, vigorous changes will be necessary; anything less than a reduction by a factor of two is unlikely to give a significant improvement in reaction.

From Chapter 3: design for dynamic loading – initial design

This important section contains a number of useful suggestions – in view of its length it is not repeated here but well worth revisiting.

From Chapter 4: specific dynamic loadings

(a) Wind-induced vibrations comprise two broadly distinguishable mechanisms – gusts and aerodynamic instabilities. Wind gust actions are usually treated via a power spectrum approach. Aerodynamic instabilities often exhibit two different mechanisms – either forced excitation due to instabilities in the flow pattern (e.g. vortex shedding) or there are specific structural shapes susceptible to self-excited oscillation (e.g. galloping). Examples of solutions are:
— bridge deck flutter – design out by change of shape;
— iced cable galloping – minimize by dampers and stays;
— chimney vortex shedding – design out via helical strakes.
(b) Earthquake loading is transient in nature, inertial forces tend to predominate and the nature of the ground is important, as is the proximity to faults. The response spectrum approach is commonly used, although a number of other approaches (mostly variants) are possible. Design strategies include 'conventional', utilizing standard structural forms/provision of ductility/appropriate detailing, and 'active and passive control', which usually involves modified damping and/or base isolation (which causes a frequency shift).
(c) Walking and rhythmic activities may lead to energy input at fundamental frequencies of between (say) 1.5 and 2.8 Hz. Further energy may be input up to

the third Fourier component, i.e. up to 8.4 Hz. Rhythmic activity loading also involves consideration of contact ratio, synchronization factor and horizontal (as well as vertical) effects. For floors, those with frequencies below 7 Hz should be carefully considered as problems may arise. For footbridges with spans over 25 m similar careful consideration should also be given and supplementary damping may be advisable.

(d) Loading resulting from explosions in general comprises two distinct physical processes: a blast-wave front behind which the air pressure and density differs substantially from the initial ambient valves and the airflow velocities implicit in these density changes. Characteristically there is a sharp rise in pressure, a rapid decay after and reflection effects can be significant. In contrast gas explosion loading exhibits a relatively slower build-up and extended resulting pressure pulse for domestic gas explosions. Venting can be significant in reducing peak blast pressures. Dynamic response to blast often follows simplified elastoplastic analysis allowing for substantial inelastic deformation.

(e) Three classes of machinery are considered, namely rotating, reciprocating and impacting. Typical excitation for rotating/reciprocating machinery occurs at 10 to 50 Hz, due to out of balance masses and/or gas forces. Isolation mountings to reduce vibration levels are not uncommon. Design is usually carried out to avoid resonance at normal operating speeds. 'Low tuned' foundations may be necessary in view of the difficulty of achieving structural frequencies in excess of 50 Hz. Dynamic forces are usually dominated by the basic frequency component but a range of higher frequency components may also be significant.

(f) Ground-transmitted vibration (comprises) a number of elastic waves (including) body waves, shear waves and surface waves. A commonly used characterization of ground motion is the peak particle velocity (PPV). Vibration standards include BS6472 (human) and BS7385 (structural). Many problems arising from man-made vibration relate to human response, affected both by structural motion and air motion i.e. acoustics.

(g) Impact loading can range from high velocity, low mass (e.g. missiles) to low velocity, high mass (e.g. ship impact on bridge piers). Inelastic and plastic stress waves occur during the 'early time' response. Overall structural dynamic response usually follows later and can often be decoupled from the 'early time' response. The requirement may be for the structure to withstand the impact without serious damage. A quasi-static design approach is used extensively in civil and structural engineering.

(h) Hydrodynamic loading principles parallel those of wind action. For wave excitation of structures in deep water, for example an oil exploration platform, where predominant wave excitation may be in the range 0.07 to 0.20 Hz, structural design is to avoid resonances in the range 0.03 to 0.25 Hz. Response spectrum analysis for fixed structures is not uncommon. For current excitation of piers/columns in shallow water there are possibilities of vortex shedding. It is usual to ensure that resonance can not occur by ensuring that the critical speed for the lowest mode is greater than the maximum practical current speed.

Illustrative Example 1: response of buildings to a gusty wind

The following two-part example considers buildings which are representative of recent UK structural practice, but which attracted adverse comment about dynamic response, provoking detailed studies.

Part 1 – tower block

The first building considered is a tower block 80 m high in an inland urban location (Ellis et al., 1979). The design provision for lateral forces was based on two service cores of in-situ concrete. In practice, the total lateral stiffness was found greatly to exceed the calculated stiffness of the cores, which comprised a number of separate wall sections as a consequence of the need for openings. The actual stiffness was derived partly by complex interaction of core and peripheral columns through the deep-slab floors, but it was concluded that about half the total was derived from notionally non-structural partitions. This gave natural frequencies of 0.68 Hz north–south, 0.86 Hz east–west and 0.79 Hz in torsion. The rule-of-thumb frequency estimate $46/H$ gives 0.58 Hz (BS6399 Part 2 baseline $60/H = 0.75$ Hz). The outline is a plain rectangular prism $36 \text{ m} \times 20 \text{ m}$ in plan (long dimension east–west).

The dominant complaint from the owner was of on-going, repeated damage to plaster finishes to the extensive internal partitioning. Monitoring gave emphasis to winds in a direction making about 30° with the east–west axis of the building, when accelerations in the north–south mode were about twice the concurrent values east–west.

To use the 'lattice plate' extension of the line-like model of gust action as a simple predictor of dynamic sensitivity, the north–south response is addressed as the alongwind response to a south wind. Using the observed natural frequency and damping (0.9% of critical damping, log. dec. 0.05 at the small amplitudes in question), the prediction of the 'narrow band' dynamic response proves accurate within the limits of the assumptions that have to be made on the various wind input parameters. The mean, RMS broad-band (large gust, quasi-static) and RMS narrow-band (quasi-resonant) components of displacement (broadly proportional to induced stresses) are in the ratios $1 : 0.36 : 0.15$ (calculation on pages 64–65).

These ratios are typical of this class of structure and are compatible with the 6% dynamic augmentation of stress given in BS6399 Part 2 (factor C_r) if the structure was classified as an office building *without* additional masonry subdivision. Although the blockwork partitions in this building raise the possibility of presuming the more favourable value $K_b = 0.5$ ($C_r = 0.03$), it can be noted that this presumes grossly higher damping (6.2% of critical damping, equation C7 of BS6399 Part 2); this might be appropriate at the ULS level but it is certainly not so for serviceability checks.

This study is thus inconclusive on the role of dynamic response in the observed damage to finishes. The dynamic properties, notably the natural frequencies, were not inferior to the norms and the dynamic response thus not worse than the norm. On a metal structure this level of dynamic response would have a greater fatigue effect than the large-gust quasi-static stress fluctuations, although still rarely a significant design influence. In the absence of information on damage to partition finishes under small cyclic loads, it can only be guessed that there may be some incremental crack-opening mechanism. The measured natural frequencies show that the partitions participated strongly in the lateral resistance of the structure; without that participation, the natural frequencies would not have achieved the 'norm' values and dynamic response

Design Guide: Illustrative Example 1	Made by: TAW	Sheet 1 of 2
Dynamic Gust Response of a Building	Checked by: ICE	Date: 1999

Prismatic building as shown:

Measured properties (Ellis *et al.*, 1979)
Lowest natural frequency 0.68 Hz
Mode shape, closely $\phi_1 = z/H$ (independent of y)
Modal generalized mass $M_1 = 6000$ t
Natural damping $\delta_s = 0.05$

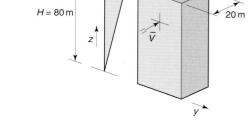

Location 100 km from sea, 3 km town.
BS6399 Part 2 (Tables 22 & 23)
$S_c = 1.33 \quad S_t = 0.145 \quad T_c = 0.92 \quad T_t = 1.22$

Response measurements published for
basic wind speed $V_b = 12$ m/s.
For initial assessment, enter 'simple' formulation
with values of wind parameters at $z_R = \tfrac{3}{4}H = 60$ m,
i.e. $\bar{V}_{ZR} = 12 \times 1.33 \times 0.92 = 14.7$ m/s (kinematic pressure $\bar{q} = \tfrac{1}{2}\rho\bar{V}^2 = 0.132$ kN/m²),
$\sigma(v) = 14.7 \times 0.145 \times 1.22 = 2.6$ m/s

The roughness change sea-country is effectively complete; the change country-town has generated about one-half of the equilibrium value of T_t (1.00 at edge of town, 1.41 equilibrium value) so the spectral time scale will be taken as average of 'town' and 'country' values, i.e. $T_s = 12$ s (a low estimate is conservative).
Thus the resonant spectral ordinate
$n_1 S(n_1) = 0.6\sigma^2(v)/(11.9 n_1 T_s)^{2/3} = 0.6 \times 2.6^2/(11.9 \times 0.68 \times 12)^{2/3} = 0.19$ (m/s)²

The aerodynamic admittance for a lattice plate is evaluated as $J^2 = J_z^2 \times J_y^2$, where J_z^2 is the value for a line structure $\phi_z = z/H$ with $H = 80$ m and J_y^2 is the value for a line of 0.6 times the actual length of the smaller dimension, i.e. $h = 0.6 \times 36 = 21.6$ m (with $\phi_y = 1$). For this 'bluff body' the scale parameter of the normalized co-spectrum of pressure is taken as $c = 6$. Thus

$$h_{tez} = \frac{cn_1}{\bar{V}} \frac{(\int \phi_z \, dz)^2}{\int \phi_z^2 \, dz} = \frac{6 \times 0.68}{14.7} \frac{(80/2)^2}{80/3} = 16.6 \qquad J_z^2 = \frac{2}{h_{tez}}\left(1 - \frac{1}{h_{tez}}\right) = 0.113$$

$$h_{tey} = \frac{6 \times 0.68}{14.7} 21.6 = 60 \qquad J_y^2 = 0.28$$

so $J^2 = 0.031$, $J = 0.18$.

The aerodynamic damping is negligible (see next page), so $\left\{\dfrac{\pi^2}{2\delta}\right\}^{1/2} = 10$

Thus $\quad \dfrac{\sigma_N(Y)}{\bar{Y}} = \left\{\dfrac{\pi^2}{2\delta}\right\}^{1/2} 2J\left\{\dfrac{n_1 S(n_1)}{\bar{V}^2}\right\}^{1/2} = 10 \times 2 \times 0.18 \times \dfrac{(0.19)^{1/2}}{14.7} = 0.11$

Design Guide: Illustrative Example 1	Made by: TAW	Sheet 2 of 2
Dynamic Gust Response of a Building	Checked by: ICE	Date: 1999

The drag coefficient is taken as $C_D = 1.1$

The aerodynamic damping at this speed is negligible: mean load per unit area $\bar{p} = 0.132 \times 1.1 = 0.15$ N/m². If modal mass is 6000 t, the actual mass per unit area of face $m \approx 6$ t/m². $\bar{V} = 14.7$ m/s, $n_1 = 0.68$ Hz
Thus $\delta_a = \bar{p}/mn\bar{V} = 0.002$

The mean value of the first-mode deflection is

$$\bar{Y}_1 = \frac{\bar{P}_1}{K_1} = \frac{1/2 \times 1.1 \times 0.132 \times 80 \times 36}{6000 \times (2\pi \times 0.68)^2} = 0.0019 \text{ m} = 1.9 \text{ mm}$$

The accelerations derive from the narrow band response

$$\sigma(\ddot{Y}) = (2\pi n_1)^2 \sigma_N(Y_1) = (2\pi \times 0.68)^2 \times 0.11 \times 1.9 = 4 \text{ mm/s}^2$$

The observed acceleration response when the wind was roughly perpendicular to the face was typically close to this value, but values nearly twice as large were observed in oblique winds.

<u>5 year check</u>

The wind speed for return period equal to 5 years is about $V_b = 19$ m/s. $h_{t_{ez}}$ and $h_{t_{ey}}$ are both increased pro-rata, which increases the aerodynamic admittance. Thus $J_z^2 = 0.172$, $J_y^2 = 0.39$, giving $J^2 = 0.067$. The change in T_s will be ignored. $\sigma_N(Y)/\bar{Y}$ is thus increased by ×1.4. The mean value is increased $\propto \bar{V}^2$, i.e. by factor 2.5.

The predicted RMS acceleration is thus $4 \times 1.4 \times 2.5 = 14$ mm/s², considering mode 1 in 'face on' wind. If the actual values are increased pro-rata, this would give 25 mm/s². There will also be contributions from the torsional mode affecting locations near the corner of the building. However, as the 'acceptable' limit (Lawson 1982 etc) at 0.68 Hz is about 50 mm/s² (0.050 m/s²) (Figure 9), adverse subjective response is unlikely.

<u>50 year check</u>

The 50 year check return wind-speed is 1.16 times the 5-year value (BS6399 Part 2 Appendix D). Repetition of the above procedure gives $J = 0.29$ or 1.6 times the value in the calculation for $V_b = 12$ m/s. Thus

$$\sigma_N(Y_1)/\bar{Y}_1 = 0.18$$

The BS6399 Part 2 design value $V_e = \bar{V}(1 + g_t S_t T_t) = 1.44\bar{V}$ (with $g_t = 2.5$). Thus for a load-effect f (for example, base shear) $f_e = 2.08f$.

To assess the dynamic augmentation of f_i note that in a cycle of dynamic response with RMS $\sigma(f)$, the maximum $\tilde{f} = \sqrt{2}\sigma(f)$; this is a common-sense estimate of the addition to the maximum static response. As base shear will be closely proportional to Y_1, the augmentation is $0.18\sqrt{2}/2.08$, or 12%.

would have been greater. It is more important to consider the balance and distribution of elements contributing resistance to lateral forces than to target arbitrary totals.

Part 2 – slab block   The second building considered is a relatively modest slab block standing 27 m high above an extensive podium, in an exposed near-coastal location (Wyatt and Best, 1984). The plan dimensions are 40 m × 12 m, and the design provision for lateral forces perpendicular to the long face comprises two simple and effective full-depth shear walls 27 m apart. The stiffness in this case proved well-modelled theoretically, giving a natural frequency of 2.8 Hz (cf. 46/H = 1.7 Hz, 60/H = 2.2 Hz). It is likely that no explicit design check had been made for longitudinal wind loading, where resistance depends on the mullions in precast cladding panels to achieve 2.1 Hz. The torsional frequency is 3.2 Hz. The glazing is at the inside edge of the deep mullions, and the occupiers' complaints included excessive wind noise as well as perception of sway.

Monitoring confirmed the subjective reaction of the occupants. Extrapolation to the six-year return period wind speed of 26 m/s at roof level suggested response of about ten times the median threshold of perception of motion at this frequency, roughly twice the recommended limits, which are discussed in greater detail in the reference cited above. However, the subsequent history of this building is that complaints did not persist, given familiarity and reassurance that perceived motion was not an indication of structural inadequacy. It should also be noted that the biggest contribution to dynamic response came from the torsion mode, so that maximum values were restricted to a small fraction of the occupied space near the corners of the building on the upper floors, and the ends of the building (i.e. beyond the shear walls) were used only for services and stair towers.

As in the previous example, the maximum observed response occurred in winds obliquely incident to the long face. Lattice–plate analysis of downwind and torsional responses under wind perpendicular to the long face proved a useful indication, although the equilibrium boundary layer model (i.e. wind structure appropriate to a long approach over terrain of uniform roughness, for which standard values are relatively well established) probably underestimates the actual response. This building was located some 2 km downwind of an increase of terrain roughness at the coast, from which the mean speed at any given height decreases rather slowly but the gustiness increases more rapidly, as can be seen by comparing factors S_c (mean) and S_t (relative gustiness) in Table 22 of BS6399 Part 2. Although the total ULS load effect decreases from the value applicable at a location of uniform lower roughness, the dynamic contribution may actually increase in this transition zone.

The dominant reason for the high dynamic response of this simple and stiff structure, however, is its geographical location on the north-west edge of the British Isles, giving very high values of wind speed at the relevant frequencies of occurrence for subjective reaction assessments. Significant additional costs would have been incurred if this problem had been included in the design process with the objective of meeting the response criteria. The structure as built was already highly effective; the simplest measure would have been to increase torsional resistance by moving the shear walls to the end of the building, but this would not have been compatible with the existing concept of lightweight and translucent staircase towers.

This two-part illustrative example suggests that many buildings lie close to the threshold at which dynamic response would need to be a serious design consideration, but that the threshold is 'soft', not a firm limit in the manner of strength criteria. The crude analysis used here gives some guidance as a sensitivity study, and suggests that height *per se* is not a dominant factor, although slenderness is important and it is clearly a broad generalization that taller structures are likely to be more slender. The greatest sensitivity is to the basic wind speed as a function of geographical location. Some structural input factors may be difficult to estimate for checks where moderate stress levels are at issue; in such cases much lower values of damping should be assumed than the values implicit in BS6399 Part 2 factor K_b which are directed to an ULS check.

Dynamics: an introduction for civil and structural engineers

Illustrative Example 2: response of a building to an earthquake

The following example considers a 20-storey office/residential building, founded on a highly soft soil and situated in UBC Zone 4 in the USA, which represents a relatively active seismic area. The principal lateral force resisting structure is a steel special moment resisting frame. An equivalent horizontal seismic force of 1014 kN, representing nearly 6% of the effective vertical load, is shown to be required to be resisted, according UBC (1997) (see the calculation sheets below).

Design Guide: Illustrative Example 2	Made by: EDB	Sheet 1 of 4
Seismic analysis of building to UBC:97	Checked by: ICE	Date: 1999

Given data

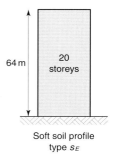

Soft soil profile type s_E

UBC Zone 4
Building function: office/residential
Floor loadings (assumed equal on all floors)
Dead = 750 kN/floor
Partitions = 75 kN/floor
Live = 450 kN/floor
Lateral force resisting structure – steel special moment resisting frame.

Analysis to UBC (1997) Chapter 16 Division IV

Total seismic base shear V

$\quad V = (C_V I)/(RT) W$ Eqn 30–4
but $\quad V \leq (2.5 C_a I/R) W$ Eqn 30–5
and $\quad V \geq (0.11 C_a I) W$ or Eqn 30–6
$\quad \geq (0.8 Z N_v I/R) W$ if greater Eqn 30–7
(Zone 4 only)

Evaluate these parameters from UBC tables.

$Z = 0.4$	(for Zone 4)	Table 16-I
$C_v = 0.96\ N_v$	(for $Z=0.4$, soil type s_E)	Table 16-R
$C_a = 0.36\ N_a$	(for $Z=0.4$, soil type s_E)	Table 16-Q
$I = 1.0$	(for Standard occupancy – Category 4)	Table 16-K
$R = 8.5$	(for a steel special moment resisting frame)	Table 16-N

[NB = Steel frame must satisfy special detailing requirements of UBC (1997) Chapter 22 for special moment resisting frames]

$\quad T = 0.0853 \times 64^{3/4} = 1.93\,\text{s}$ (for 64 m high steel frame building) Eqn 30–8

[NB: Somewhat more favourable results may be obtained by calculating T analytically – but note restrictions on the advantage allowed in 1630.2.2.(2)].

Design Guide: Illustrative Example 2	Made by: EDB	Sheet 2 of 4
Seismic analysis of building to UBC:97	Checked by: ICE	Date: 1999

$N_a = N_v = 1.0$ (assume no active fault within 15 km of the building)   Tables 16-S 16-T

$W = ($ 750 $+$ 75 $+$ 0 $) \times$ 21   Clause 1630.1.1
          Dead load   Partitions   Applicable   No. of
                              Live load   floors & roof

$= 17\,325$ kN

Hence subject to
$V = (0.96 \times 1.0 \times 1.0) / (8.5 \times 1.93)\, W = 5.85\%\, W$ Eqn 30–4
$V \leq ((2.5 \times 0.36 \times 1.0 \times 1.0)/8.5)\, W = 10.59\%\, W$ Eqn 30–5
$\geq (0.11 \times 0.36 \times 1.0)\, W = 3.93\%\, W$ Eqn 30–6
$\geq ((0.8 \times 0.4 \times 1.0 \times 1.0/8.5)\, W = 3.76\%\, W$ Eqn 30–7

Eqn 30–4 governs and

$V = 5.85\%\, W = 1014$ kN

From Clause 1629.8.3 number (2), an equivalent static analysis is permitted, provided no irregularities exist in plan or elevation.

Vertical distribution of forces

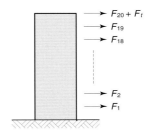

$V = F_t + \sum_{i=1}^{N} F_i$ Eqn 30–13

$F_t = 0.07\, TV$ but is not greater than $0.25V$

and $F_t = 0$ for $T \leq 0.7$ s   Eqn 30–14

Therefore $F_t = 0.07 \times 1.93\, V = 0.135\, V = 137$ kN

$$F_x = \frac{(V - F_t) w_x h_x}{\sum_{i=1}^{N} w_i h_i}$$ Eqn 30–15

where w_x and h_x are the weight and height at level x

Design Guide: Illustrative Example 2	Made by: EDB	Sheet 3 of 4
Seismic analysis of building to UBC:97	Checked by: ICE	Date: 1999

This is evaluated on a spreadsheet

Storey number, i	Height of storey h_i (m)	Weight of floor w_i (kN)	$w_i h_i$ (kN m)	Storey lateral force (kN)
20	64	825	54 400	220.5
19	60.8	825	51 680	79.3
18	57.6	825	48 960	75.2
17	54.4	825	46 240	71.0
16	51.2	825	43 520	66.8
15	48	825	40 800	62.6
14	44.8	825	38 080	58.5
13	44.8	825	35 360	54.3
12	41.6	825	32 640	50.1
11	38.4	825	29 920	45.9
10	35.2	825	27 200	41.8
9	32	825	24 480	37.6
8	28.2	825	21 760	33.4
7	25.6	825	19 040	29.2
6	22.4	825	16 320	25.1
5	19.2	825	13 600	20.9
4	16	825	10 880	16.7
3	12.8	825	8160	12.5
2	9.6	825	5440	8.4
1	6.4	825	2720	4.2
Ground	3.2	825	0	0.0
Σ		17 325 Total seismic weight (kN)	571 200 $\Sigma w_i h_i$ (kN m)	1013.8 Total seismic base shear, $V = \Sigma(F_i + F_t)$

From Eqn 30–1, these are multiplied by ρ and applied as horizontal loads along each major axis of the building separately (but also see clause 1633.1)

ρ is the redundancy factor from Eqn 30–3 and is not less than 1.0; it must not exceed 1.25 in special frames. In most cases $\rho = 1.0$.

$P\Delta$ effects must be considered if required by 1630.1.3 but usually do not apply. $P\Delta$ clauses in Eurocode 8 appear more rational than UBC.

Design Guide: Illustrative Example 2	Made by: EDB	Sheet 4 of 4
Seismic analysis of building to UBC:97	Checked by: ICE	Date: 1999

Vertical earthquake forces are also applied at each level, equal to
$$E_v = 0.5\, C_a I D$$
where D is the dead load at that level.
$$E_v = 0.5 \times 0.36 \times 1.0 \times 750 = 135 \text{ kN per floor}$$

Clause 1630.3.1

Special rules for vertical earthquake forces apply to horizontal cantilevers and horizontal prestressed members - see clause 1630.11

The horizontal and vertical earthquake forces calculated above are combined with other forces as follows:

$$1.2\, D + 1.0\, E + (f_1 L + f_2 S) \quad \text{Eqn 12–5}$$
$$0.9\, D \pm 1.0\, E \quad \text{Eqn 12–6}$$

where
- D = Dead load
- E = Earthquake load
- L = Live load
- S = Snow load
- f_1, f_2 are reduction factors < 1.0

For example, at 10th floor level, total load applied is

Horizontal   41.8 kN (see spreadsheet on sheet 3 of 4)
Vertical $1.2 \times 750 + 1.0 \times 135 + 0.5 \times 450 + 0 = 1260$
or $0.9 \times 750 - 1.0 \times 135 = 540$
1260 or 540 kN vertically

The rules for the point of application of the horizontal load are given in clause 1630.6 and 1630.7 which allow for torsional effects.

NB: the load combination equations, 12–5 and 12–6, are modified for concrete structures – see Exception 2 of 1612.2.1.

Illustrative Example 3: floor subject to rhythmic activity

Consider a floor with a natural frequency of 6.3 Hz. Resonance is thus possible with the third Fourier component ($N = 3$) of rhythmic activity at 2.1 Hz; although somewhat higher than the most common frequency for aerobics, this frequency is possible and is within the limits specified for consideration in BS6399 Part 1. The third-harmonic steady-state response amplitude is shown in Chapter 4 to be given by

$$\tilde{y} = \frac{G}{k}\frac{\pi r_3}{\delta} \quad \text{where } r_3 = \frac{\sqrt{2(1 + \cos 6\alpha\pi)}}{36\alpha^2 - 1}$$

The vigour of the activity is expressed through the contact ratio, α; greater vigour is expressed by smaller values of α. However, the specific suggested value $\alpha = 1/2$ for high impact aerobics then gives $r_3 = 0$, because with the resulting 3 : 2 ratio of pulse duration to natural period in each cycle, the floor just comes to rest in static equilibrium at the instant the body leaves contact, and there is no resonant carry-forward to the next pulse. The maximum response during each contact phase (assuming the semi-sinusoid pulse shape) is almost independent of damping, viz., $\hat{y} = (3\pi/2)(G/k)$.

The maximum resonant response to aerobics or dancing (excluding jumping) is in fact given by less vigorous activity, trial and error giving $r_3 = 0.142$ for $\alpha = 0.63$. As this response is at 6.3 Hz whereas the non-resonant response at $\alpha = 1/2$ is dominated by the excitation frequency 2.1 Hz, the resonant case will give larger accelerations in practical cases, but the non-resonant case should be included in stress checks if the damping exceeds about 0.1 logarithmic decrement (critical damping ratio $\zeta = 0.016$). The resonant response with jumping excitation giving a lower contact ratio is potentially much more severe, up to $r_3 = 1.13$ for high jumping such that $\alpha = 1/4$; this is eight times more severe than the value $r_3 = 0.142$ identified above as the critical resonant component for aerobics.

A typical practical floor with frequency 6.3 Hz might comprise a concrete ribbed slab spanning $L = 9$ m and $m_s = 500$ kg/m^2. The mode shape is thus $\mu = \sin \pi x/L$ and the modal generalized mass $M = \int m_s \mu^2 \, dA$ (where dA is an element of the floor area) becomes $M = \frac{1}{2}m_s L b$, where b is the breadth of the floor. Using the notation $\varpi_0 = 2\pi n_0 = 2\pi/T_0$, the corresponding generalized stiffness is $K = M\varpi_0^2$. The modal generalized force $P = \int p\mu \, dA$, in which p is the actual load per unit area, is likewise $P = (2/\pi)pLb$.

Incorporating the above in SDOF analysis for steady-state response, defining G as the exciting weight (effective weight of people engaged) per unit area, the modal response amplitude is

$$Y = \frac{\pi r_3}{\delta}\left(\frac{2}{\pi}GLb\right) / \left(\frac{1}{2}m_s L b \varpi_0^2\right) = \frac{4Gr_3}{\delta m_s \omega_0^2}$$

and the acceleration amplitude at mid-span where $\mu = 1$ is

$$\tilde{a} = \varpi_0^2 Y = 4Gr_3/\delta m_s$$

The suggested allowance for lack of correlation of the motion of a number of persons by multiplication by 2/3 would be applicable to this result.

At the frequency in question the BS6472 'curve number' for subjective comfort assessment is 200 times the RMS acceleration expressed in m/s^2, equivalent for harmonic motion to 140 times the amplitude. As floors used for dancing or aerobics are likely to be simple and sparsely furnished, logarithmic decrement $\delta = 0.15$ (critical damping ratio $\zeta = 0.024$) may be more appropriate than the values up to $\zeta = 0.03$ suggested for domestic or office floors (SCI-P-076). Including the correlation factor, the BS6472 curve number R (say), is thus

$$R = 2500 G r_3 / m_3 \quad \text{(units of N, kg, or kN, t)}$$

Suggested values for G for serviceability limit states have included 0.4 –0.6 kN/m² for aerobics and dancing, respectively, but up to 1.5 kN/m² for crowd activities. Thus, for aerobics the suggested example gives $R = 2500 \times 0.4 \times 0.142/0.50 = 280$. This would be perceived as quickly tiring and might cause damage to finishes, but would be unlikely to cause panic. It would also pose a serious risk of leading to unacceptable perception of vibration in adjoining areas where the occupants would be more critical, such as spectator areas. For jumping, however, the suggested four-fold increase of crowd density and eight-fold increase in factor r_3 would clearly render a floor of this type unacceptable.

The mid-span bending moment modal influence factor (moment per unit width per unit modal displacement) for this simple span would be $\pi^2 EI/L^2$, and as $\varpi_0^2 = \pi^4 EI/m_s L^4$, the moment per unit width (**M**, say) derived from the resonant part of the response is

$$\mathbf{M} = \frac{4 G r_3}{\delta m_s} \frac{m_s L^4}{\pi^4 EI} \frac{\pi^2 EI}{L^2} = \frac{4 G r_3 L^2}{\pi^2 \delta}$$

Comparing this with the static value $\mathcal{M}_0 = G_0 L^2/8$, where G_0 is the design static imposed loading and with correlation factor 2/3

$$\frac{\mathbf{M}}{\mathbf{M}_0} = \frac{2}{3} \frac{4 G r_3 L^2}{\pi^2 \delta} \frac{64 r_3}{3\pi^2 \delta} \frac{G}{G_0} = \frac{2}{3} \frac{4 G r_3 L^2}{\pi^2 \delta} \frac{8}{G_0 L^2} = \frac{64 r_3}{3\pi^2 \delta} \frac{G}{G_0}$$

There is little existing guidance on the interpretation of this result. For aerobics or dancing there is no problem within the given range of G but the high jumping with $r_3 = 1.13$, $G = 1.5$ kN/m² and $G_0 = 5$ kN/m², say, leads to $\mathbf{M}/\mathbf{M}_0 = 0.73/\delta$. Even with recognition that it is legitimate to assume enhanced damping in a ULS check, this is clearly not likely to be acceptable, reinforcing the view that floors with a natural frequency permitting third-harmonic resonance are not suitable for such activity.

Classified selected vibration standards (at end of 1998) (sorted into British, international and other standards)

Human	Structural	Machinery	Calibration and other

Exposure (1–80 Hz)
  BS 6472
  (ISO 2631/2)
General requirements
  ISO 2631/1, 2 and 3
Shipboard vibration data
  BS 6632 and 6633
  (ISO 4867 and 4868)
Hand transmitted vibration
  ISO 5349
Effect of vibrations on people
  VDI 2057
	Buildings
	  BS 7385: Part 1
	  (ISO 4866)
	  BS 7385: Part 2
	  DIN 4150/1,/2 & /3
	Machine foundations
	  BS CP 2012:Part 1
	  DIN 4024/1 & 2
	Shipboard vibration data
	  BS 6632 and 6633
	  (ISO 4867 and 4868)
	Merchant ships
	  BS 6634
	  ISO 6954
	Vibration in railway tunnels
	  ISO 10815
	Steel stacks
	  DIN 4133
	Vibration serviceability
	  ISO/DIS (Draft) 10137
		Rotating electrical machines
		  BS 4999
		Balance quality – rigid rotors
		  ISO 1940/1
		Mechanical vibration – Instruments
		  ISO 2954
		Large rotating machines
		  ISO 3945 (superseded by ISO 10816)
		Non-reciprocating machines
		  ISO 7919/1, 2, 3, 4 and 5
		Reciprocating machines – generator sets
		  ISO 8528–9
		Sensitivity to unbalance
		  ISO 10814
		Mechanical vibration – Evaluation
		  ISO 10816 (**supersedes ISO 2372**)
		  Part 1 – general
		  (**supersedes ISO 3945**)
		  Part 2 – steam turbines
		  Part 3 – industrial machines
		  Part 4 – gas turbines
		  Part 5 – hydraulic turbines
		  Part 6 – reciprocating machines
		  ISO 10817 (to be published)
		Rotating shafts
		  VDI 2056
		Effect of vibrations on machines
			Vibration and shock pickups
			  BS 6955
			  (ISO 5347)
			Mechanical mounting of accelerometers
			  BS 7129
			Instrumentation
			  ISO 8041
			Noise & vibration control
			  BS 5228
			  Part 1 – basic information
			  Part 2 – demolition
			  Part 3 – open cast mining
			  Part 4 – piling
			  Part 5 – surface mineral extraction (not open cast)

Bibliographic sources

Bibliography related to human exposure	**Bibliography related to structures/equipment**	**Bibliography related to machines**	**Bibliography related to condition monitoring**

ISO 8662
  Hand held power tools
	ISO 8569
	  Sensitive equipment in buildings
		API 610
		  Centrifugal pumps
		API 612
		  Steam turbines
		API 613
		  Gear units
		API 617
		  Centrifugal compressors
		API 618
		  Reciprocating compressors
		API 670
		  Monitoring systems
			Machines
			  ISO/WD 13380–1.2
			  Date:1997–09–14
			Terminology (to be published)
			  ISO 13372
			Instrumentation (to be published)
			  ISO 13377
			Transducers (to be published)
			  ISO 13378

References combined with selected codes and standards

AASHTO Guide Specifications for Seismic Isolation Design. American Association of State Highway and Transportation Officials (AASHTO), Washington, DC, 1991.

Abé, M. and Fujino, Y. Dynamic characteristics of multiple tuned-mass dampers and some design formulas. *Earthquake Engineering and Structural Dynamics*, **23**, 813–835, 1994.

Al-Hassani, S.T.S. and Reid, S.R. The effects of high strain rates on material properties. Report OTI 92 602. HSE Offshore Technology Information, HMSO, London, 1992.

Arnold, R.N., Bycroft, G.N. and Warburton, G.B. Forced vibration of a body on an infinite elastic solid. *Journal of Applied Mechanics (ASME)*, **22**, 391–400, 1955.

American Concrete Institute. ACI 3007-95 Standard Practice for the Design and Construction of Reinforced Concrete Chimneys. American Concrete Institute, Detroit, 1995.

American Society of Civil Engineers. *Seismic Analysis of Safety-related Nuclear Structures*. Standard 4-86, ASCE, New York, 1986.

Applied Technology Council, *Seismic Evaluation and Retrofit of Concrete Buildings*. Applied Technology Council, Redwood City, CA, 1996.

Barkan, D.D. *Dynamics of Bases and Foundations*. McGraw-Hill, New York, 1962.

Barltrop, N.D.P. and Adams, A.S. *Dynamics of Fixed Marine Structures (UR8)*, 3rd edition. Butterworth-Heinemann, Oxford, 1990.

Biggs, J.M. *Introduction to Structural Dynamics*. McGraw-Hill, New York, 1964.

Blake, L.S. *Civil Engineer's Reference Book*, 3rd edition. Butterworths, London, 1975.

Blevins, R.D. *Flow Induced Vibrations*, 2nd edition. Van Nostrand Reinhold, New York, 1977.

Blevins, R.D. *Formulas for Natural Frequency and Mode Shape*. Van Nostrand Reinhold, New York, 1979.

Bolt, B. *From Earthquake Acceleration to Seismic Displacement*. Fifth Mallet–Milne Lecture. SECED/John Wiley, 1995.

Bolton, A. *Structural Dynamics in Practice*. McGraw-Hill, 1994.

Booth, E. (editor). *Concrete Structures in Earthquake Regions*. Longmans, Harlow, 1994.

Brebbia, C.A. and Walker, S. *Dynamic Analysis of Offshore Structures*. Butterworths, London, 1979.

Brown, C.W. *Dynamic Behaviour of Bridges*. TRRL Supplement Report 275 (paper 8), 1977.

BSCP 2012/1. *Code of Practice for Foundations for Machinery: Foundations for Reciprocating Machines*. British Standard Code of Practice BSCP 2012, Part 1, 1974.

BS4485. Water Cooling Towers: Part 4: Code of Practice for structural design and construction. British Standards Institution, 1996.

BS5228. Noise control on construction and open sites; Part 4: Code of practice for noise and vibration control applicable to piling operations (AMD 7787). British Standards Institution, 1992.

BS5400. Steel, concrete and composite bridges: Part 1: General statement, British Standards Institution, 1988.

BS6399. Loading for buildings. Part 2: Code of practice for wind loads, British Standards Institution, 1997.

BS6472. Guide to evaluation of human exposure to vibration in buildings (1 Hz to 80 Hz). British Standards Institution, 1984.

BS7385. Evaluation and measurement for vibrations in buildings. Part 1: Guide for measurement of vibrations and evaluation of their effects on buildings. British Standards Institution, 1990.

BS8100. Lattice towers and masts. Part 2: Guide to the background and use of Part 1 code of practice for loading. British Standards Institution, 1986.

BS8110. Structural use of concrete. Part 1: Code of practice for design and construction. British Standards Institution, 1997.

Bullen, K.E. and Bolt, B.A. *An Introduction to the Theory of Seismology*, 4th edition. Cambridge University Press, Cambridge, 1995.

Burdekin, M. (editor). *Seismic Design of Steel Structures after Northridge and Kobe*. SECED/Institution of Structural Engineers, 1996.

Carney, III, J.F. Motorway impact attenuation devices: past, present and future. In: *Structural Crashworthiness and Failure*, Jones N. and Wierzbicki, T. (editors). Elsevier Applied Science, Oxford, 423–466, 1993.

Casciati, F. Active control of structures in European seismic areas. *Proceedings of the Eleventh World Conference on Earthquake Engineering*. Pergamon Press, Oxford, 1996.

CEN 94. CEN Eurocode 1 part 2-4. Wind actions. ENV 1991-2-4, Brussels, 1994.

Chen, P.W. and Robertson, L.E. Human preception thresholds of horizontal motion. *Proceedings of the American Society of Civil Engineers*. **98**, ST8, 1681–1696, 1972.

Chopra, A.K. *Dynamics of Structures: Theory and Applications to Earthquake Engineering*. Prentice-Hall, Englewood Cliffs, NJ, 1995.

Clough, R.W. and Penzien, J. *Dynamics of Structures*. McGraw-Hill, New York, 1975.

Clough, R.W. and Penzien, J. *Dynamics of Structures*. McGraw-Hill, New York, 1993.

Collins, J.A. *Failure of Materials in Mechanical Design*, 2nd edition, 1993.

Corbett, G.G., Reid, S.R. and Johnson, W. Impact loading of plates and shells by free-flying projectiles: a review. *International Journal of Impact Engineering* **18**, 141–230, 1996.

Canadian Standards Association. *Steel Structures for Buildings: Limit State Design*. CAN 3 S161.1, M84 (Appendix G). Canadian Standards Association, Rexdale, Ontario, 1964.

Davenport, A.G. The application of statistical concepts to the wind loading of structures. *Proceedings of the Institution of Civil Engineers* **19**, 447–472, 1961.

Davenport, A.G. The response of slender, line like structures to a gusty wind. *Proceedings of the Institution of Civil Engineers* **23**, 389–407, 1964.

Davenport, A.G. A note on the distribution of the largest value of a random function *Proceedings of the Institution of Civil Engineers* **28**, 187–196, 1964.

Department of Transport. *Design Rules for Aerodynamic Effects on Bridges*, BD49/93. DOE/DOT Publications, London, 1993.

Dowrick D. *Earthquake Resistant Design for Engineers and Architects*, 2nd edition. John Wiley, Chichester, 1987.

Dyrbye, C. and Hansen, S.O. *Wind Loads on Structures*. John Wiley, Chichester 1996.

Earthquake Engineering Research Institute. *Earthquake Basics Brief No. 1. Liquefaction: What It Is And What To Do About It*. EERI, Oakland, CA, 1994.

Earthquake Spectra. Theme issue: passive energy dissipation. *Earthquake Spectra* **9**, No. 3, 1993.

Ellis, B.R. and Crowhurst, D. The response of several LPS maisonettes to small gas explosions. IStructE/BRE Seminar: *Structural design for Hazardous Loads: The Role of Physical Tests*, 1991.

Ellis, B.R., Evans R.A., Jeary, A.P. and Lee, B.E. The wind induced vibration of a tall building. *Proceedings of the IAHR/IUTAM Symposium on Practical Experiences with Flow Induced Vibrations*. Karlsruhe, III: 63–69, 1979.

Ejiri, J. and Goto, Y. Introduction of Topographical Effects on Site Response for Design Spectra. *Proceedings of the Eleventh World Conference on Earthquake Engineering*. Pergamon Press, Oxford, 1996.

ESDU Data Item 85038, Revision A. *Circular–Cylindrical Structures Dynamic Response to Vortex Shedding; Part 1. Calculation Proceedures and Derivations*. ESDU International, London, 1986.

ESDU Data Item 85039, Revision A. *Circular-Cylindrical Structures Dynamic Response to Vortex Shedding; Part 2. Simplified Calculation Proceedures*. ESDU International, London, 1986.

References combined with selected codes and standards

ESDU Data Item 85020, revision E. *Characteristics of Atmospheric Turbulence Near the Ground; Part 2. Single Point Data for Strong Winds (Neutral Atmosphere)*. ESDU International, London, 1991.

ENV 1998. *Design Provisions for Earthquake Resistance of Structures*. CEN (European Centre for Standardisation), Brussels, (various dates). ENV 1991-8.

Eurocode 8. *Part 5: Foundations, Retaining Structures and Geotechnical Aspects. Annex B: Empirical Charts for Simplified Liquefaction Analyses. List of Related Codes and Standards.*

Fafjar, P. and Krawinkler, H. Sesmic design methodologies for the next generation of codes. In: *Seismic Design Practice into the Next Century: Sixth SECED Conference*, Booth, E. (editor). Balkema, Rotterdam, 1998.

Federal Emergency Management Agency. *NEHRP Guidelines for the Seismic Rehabilitation of Buildings* (FEMA 273). Federal Emergency Management Agency, Washington, DC, 1997. (See also FEMA 274 – commentary to FEMA 273.)

Federal Emergency Management Agency. *Interim Guidelines: Evaluation, Repair, Modification and Design of Welded Steel Moment Frame Structures*. Federal Emergency Management Agency, August, 1995.

Fung, Y.C. *The Theory of Aeroelasticity*. John Wiley, New York, 1955.

Goldsmith, W. Review: Non-ideal projectile impact on targets. *International Journal of Impact Engineering* **22**, 1999.

Gupta, A.K. *Response Spectrum Method in Seismic Analysis and Design of Structures*. Blackwell, Oxford, 1990.

Harris, R.I. and Deaves, D.M. *The Structure of Strong Winds: Wind Engineering in the Eighties*. CIRIA, London, Chapter 4, 1981.

Hay, J. *The Response of Bridges to Wind*. HMSO, London, 1992.

Hisada, Y. and Yamamoto, S. One-, two-, and three-dimensional effects in sediment-filled basins. *Proceedings of the Eleventh World Conference on Earthquake Engineering*. Pergamon Press, Oxford, 1996.

Hitchings, D. (editor). *A Finite Element Dynamics Primer*. NAFEMS, London, 1992.

International Association for Earthquake Engineering. *Earthquake Resistant Regulations: A World List*. IAEA, 1996.

Irwin, A.W. Human response to dynamic motion of structures. *The Structural Engineer*, **56A**, 1978.

Irwin, H.P.A.H. Cross spectra of turbulence velocities in isotropic turbulence. *Boundary Layer Meteorology* **16**, 237–243, 1979.

ISSMFE. *Technical Committee for Earthquake Geotechnical Engineering, TC4*. ISSMFE, 1993.

Irvine, M. *Structural Dynamics for the Practicing Engineer*. Allen & Unwin, London, 1986.

ISO 10137. ISO/DIS 10137, *Bases for Design of Structures – Serviceability of Buildings against Vibration*. Draft ISO/DIS 10137, International Standards Organization, Geneva, 1991.

ISO 1940-1. *Mechanical Vibration – Balance Quality Requirements of Rigid Rotors – Determination of Permissible Residual Unbalance*. ISO 1940, Part 1, International Standards Organization, Geneva, 1989.

ISO 2631-1. *Mechanical vibration and Shock – Evaluation of Human Exposure to Whole body vibration – Part 1: General Requirements*. ISO 2631, Part 1. International Standards Organization, Geneva, 1997.

ISO 2631-2. *Evaluation of human exposure to whole-body vibration – Continuous and shock induced-vibration in buildings (1 to 80 Hz)*. ISO 2631, Part 2. International Standards Organization, Geneva, 1989.

ISO 2631-3. *Evaluation of human exposure to whole body vibration – Evaluation of exposure to whole body z-axis vertical vibration in the frequency range 0.1 to 0.03 Hz*. ISO 2631, Part 3. International Standards Organization, Geneva, 1985.

ISO 2954. *Mechanical vibration or rotating and reciprocating machinery – Requirements for instruments for measuring vibration severity*. ISO 2954. International Standards Organization, Geneva, 1975.

ISO 3945. *Mechanical vibration of large rotating machines with speed range from 10 to 200 r/s – Measurement and evaluation of vibration severity in situ*. ISO 3945. International Standards Organization, Geneva, 1985.

ISO 5347-20. *Calibration of Vibration and Shock pick-ups. Primary vibration of calibration by the reciprocity method*. ISO 5347. International Standards Organization, Geneva, 1997.

ISO 5347-22. *Calibration of Vibration and Shock Pick-ups. Part 22. Acceleration Resonance Testing – General Methods.* ISO 5347, Part 22. International Standards Organization, Geneva, 1997.

ISO 5349. *Mechanical vibration – Guidelines for the measurement and the assessment of human exposure to hand-transmitted vibration.* ISO 5349. International Standards Organization, Geneva, 1986.

ISO 7919-1. *Mechanical vibration of non-reciprocating machines – Measurements on rotating shafts and evaluation. Part 1. General Guidelines.* ISO 7919, Part 1. International Standards Organization, Geneva, 1986.

ISO 7919-2. *Mechanical vibration of non-reciprocating machines – Measurements on rotating shafts and evaluation. Part 2. Large land-based steam turbine generator sets.* ISO 7919. International Standards Organization, Geneva, 1996.

ISO 7919-3. *Mechanical vibration of non-reciprocating machines – Measurements on rotating shafts and evaluation. Part 3. Coupled industrial machines.* ISO 7919. International Standards Organization, Geneva, 1996.

ISO 7919-4. *Mechanical vibration of non-reciprocating machines – Measurements on rotating shafts and evaluation. Part 4. Gas turbine sets.* ISO 7919. International Standards Organization, Geneva, 1996.

ISO 8528-9. *Reciprocating internal combustion engine driven alternating current generating sets – Measurement and evaluation of mechanical vibrations.* ISO 8528. International Standards Organization, Geneva, 1995.

ISO 10814. *Mechanical Vibration – Susceptibility and sensitivity to machines to unbalance.* ISO 10814. International Standards Organization, Geneva, 1996.

ISO 10815. *Mechanical Vibration – Measurement of Vibration Generated internally in railway tunnels by the passage of trains.* ISO 10815. International Standards Organization, Geneva, 1996.

ISO 10816-1. *Mechanical vibration – Evaluation of machine vibration by measurements on non-rotating parts – Part 1: General Guidelines.* ISO 10816. International Standards Organization, Geneva, 1996.

ISO 10816-6. *Mechanical vibration – Evaluation of machine vibration by measurements on non-rotating parts – Part 6: Reciprocating machines with power ratings above 100kW.* ISO 10816. International Standards Organization, Geneva, 1995.

Izzudin, B.A. and Smith, D.L. Response of offshore structures to explosion loading. *Proceedings of the Sixth Offshore and Polar Engineering Conference.* Los Angeles, California, 323–330, 1996.

Ji, T. and Ellis, B.R. Floor vibration induced by dance type loads. *The Structural Engineer*, 1994.

Johnson, W. and Mamalis, A.G. *Crashworthiness of Vehicles.* MEP, London, 1978.

Johnson, W. and Reid, S.R. Metallic energy dissipating systems. *Applied Mechanics Reviews*, **31**, 277–288, 1978. Updated in **39**, 315–319, 1986.

Jones, N. Bounds on the dynamic plastic behaviour of structures including transverse shear effects. *International Journal of Impact Engineering* **3**, 273–291, 1985.

Jones, N. Plastic behaviour of ship structures. *Transactions, Society of Naval Architects and Marine Engineers* **84**, 115–145, 1976.

Jones, N. *Structural Impact.* Cambridge, Cambridge University Press, 1989a.

Jones, N. On the dynamic inelastic failure of beams. In: *Structural Failure*, Wierzbicki, T. and Jones, N. editors. John Wiley, pp. 133–159, 1989b.

Jones, N. Some comments on the modelling of material properties for dynamic structural plasticity. *International Conference on the Mechanical Properties of Materials at High Rates of Strain*, Oxford. Institute of Physics Conference Series No. 102, pp. 435–445, 1989c.

Jones, N. Quasi-static analysis of structural impact damage. *Journal of Constructional Steel Research* **33**, 151–177, 1995.

Jones, N. Dynamic plastic behaviour of ship and ocean structures. *Transactions of RINA* **139A**, 65–97, 1997.

Jones, N. and Kim, S.B. A study on the large ductile deformations and perforation of mild steel plates struck by a mass. Part II: Discussion. *Transactions of the American Society of Mechanical Engineers, Journal of Pressure Vessel Technology* **119**, 185–191, 1997.

Japanese Society for Soil Mechanics and Foundation Engineering. *Manual for Zonation on Seismic Geotechnical Hazards*. Japanese Society for Soil Mechanics and Foundation Engineering, 1993.

Kabori, T. Active and hybrid structural control research in Japan. *Proceedings of the Eleventh World Conference on Earthquake Engineering*. Pergamon Press, Oxford, 1996.

Karagiozova, D. and Jones, N. Dynamic elastic–plastic buckling phenomena in a rod due to axial impact. *International Journal of Impact Engineering* **18**, 919–947, 1996.

Kelly, J.M. *Earthquake-resistant Design with Rubber*, 2nd edition. Springer Verlag, 1996.

Key, D. *Earthquake Design Practice for Buildings*. TTL, London, 1988.

Kramer, S.L. *Geotechnical Earthquake Engineering*. Prentice-Hall, Englewood Cliffs, NJ, 1995.

Lawson, M. *Design Guide to Shear Buildings*. CIRIA, London, 1982.

Malvern, L.E. The propagation of longitudinal waves of plastic deformation in a bar of material exhibiting strain-rate effect. *Journal of Applied Mechanics, (ASME)* **18,** 203–208, 1952.

Mays, C.G. and Smith, P.D. *Blast Effects on Buildings*. TTL, London, 1995.

Mazzolani, F.M. and Piluso, V. *Theory and Design of Seismic Resistant Frames*. Chapman & Hall, London, 1996.

McGuire, R. *The Practice of Earthquake Hazard Assessment*. International Association of Seismology and Physics of the Earth's Interior (IASPEI), Denver, CO, 1993.

Minorsky, V.U. An analysis of ship collisions with reference to protection of nuclear power plants. *Journal of Ship Research* **3,** No. 1, 1959.

Murray, T.M. Acceptability criterion for occupant-induced floor vibrations. *Engineering Journal AISC* **18,** 2, 1981.

Naeim F. (editor). *The Seismic Design Handbook*. Van Nostrand Reinhold, New York, 1989.

Newland, D.E., *Random Vibrations and Spectral Analysis*. Longman, London, 1975.

Newland, D.E. *An introduction to random vibration, spectral and wavelength analysis*, 3rd edition. Longman, 1993.

Newmark, N.M. and Rosenblueth, E. *Fundamentals of Earthquake Engineering*. Prentice-Hall, Englewood Cliffs, NJ, 1996.

National Research Council. *Supplement to the National Building Code of Canada, Commentary A, Serviceability Criteria for Deflections and Vibrations*. National Research Council, Ottawa, 1985.

National Research Council. *Supplement to the National Building Code of Canada, Commentary B, Wind Loads*. National Research Council, Ottawa, 1985.

Nurick, G.N. and Martin, J.B. Deformation of thin plates subjected to impulsive loading – a review. Part I: theoretical considerations; Part II: experimental studies. *International Journal of Impact Engineering* **8,** 159–186, 1989.

Paulay, T. *Simplicity and Confidence in Seismic Design*. Fourth Mallet–Milne Lecture. John Wiley, Chichester, 1993.

Paulay, T. and Priestley, M.J.N. *Seismic Design of Reinforced Concrete and Masonry Buildings*. John Wiley, Chichester, 1992.

Pappin, J.W. Design of foundation and soil structures for seismic loading. In: *Cyclic Loading of Soils*, O'Reilly, M.P. and Brown, S.F. editors. Blackie, London, pp. 306–366, 1991.

Paz, M. *International Handbook on Earthquake Engineering: Codes, Programs and Examples*. Chapman & Hall, London, 1994.

Penelis, G. and Kappos, A. *Earthquake Resistant Concrete Structures*. E.&F.N. Spon, London, 1996.

Priestley, M., Seible, F. and Calvi, M. *Seismic Design and Retrofit of Bridges*. John Wiley, New York.

Ramshaw, C.L., Selby, A.R. and Bettess, P. Computation of the transmission of waves from pile driving. In: *Ground Dynamics and Man-made Processes*, Skipp, B.O. (editor). Thomas Telford, London, 1998.

Scruton, C. and Flint, A.R. Wind-excited oscillation of structures. *Proceedings of the Institution of Civil Engineers* **27,** 673–702, 1964.

SEAOC. *Performance based seismic engineering of buildings*. Structural Engineers Association of California, Sacramento, CA, 1995.

Steel Construction Institute. *Design Guide on Vibration of Floors*. SCI-P-076. Steel Construction Institute, Ascot, 1989.

Shen, W.Q. and Jones, N. The pseudo-shakedown of beams and plates when subjected to repeated dynamic loads. *Journal of Applied Mechanics* **59,** 168–175, 1992.

Simiu, E. and Scanlan, R.H. *Wind Effects on Structures*, 2nd edition. John Wiley, New York, Chapter 13, 1986.

Skinner, R.J., Robinson, W.H. and McVerry, G.H. *An Introduction to Seismic Isolation*. John Wiley, Chichester, 1993.

Skipp, B.O. Dynamic ground movements: man-made vibration. In: *Ground Movements and their Effects on Structures*, Attwell, P.B. and Taylor, R.K. Surrey University Press, Guildford, pp. 381–434, 1984.

Skipp, B.O. Ground vibration – codes and standards. In: *Ground Dynamics and Man-made Processes*, Skipp, B.O. (editor). Thomas Telford, London, pp. 29–41, 1998.

Smith, B.W. and Wyatt, T.A. *Development of the Draft Rules for Aerodynamic Stability Bridge Aerodynamics*. TTL, London, Chapter 2, 1981.

Soong, T.T. Active research control in the US. *Proceedings of the Eleventh World Conference on Earthquake Engineering*. Pergamon Press, Oxford, 1996.

Stam, W.J. (editor). *Regulations for Type Certification of Wind Turbines*, ECN-R-005 Netherlands Energy Research Foundation, 1994.

Swiss Reinsurance. *An Atlas of Volcanism and Seismicity*. Swiss Reinsurance Company, Zurich, 1978.

Thomson, W.T. *Theory of Vibration with Applications*. Allen & Unwin, London, 1983.

Uniform Building Code. *International Conference of Building Officials*. Whittier, California, 1997.

Urlich, C.M. and Kuhlmeyer, R.L. Coupled rocking and lateral vibrations of embedded footings. *Canadian Geotechnical Journal* **10**, 145–160, 1973.

US Department of the Army. *The Design of Structures to Resist the Effects of Accidental Explosions*. Technical Manual TM5-1300. Washington DC, 1969.

Veillette, J.R. and Carney, III, J.F. Collapse of braced tubes under impact loads. *International Journal of Impact Engineering* **7**, 125–138, 1988.

Veletsos, A.S. and Wei, Y.T. Lateral and rocking vibrations of footings. *American Society of Civil Engineers* **9**, 1227, 1971.

Verweibe, C. Exciting mechanisms of rain-wind vibrations. *Structural Engineering International* **2**, 112–117, 1998.

Vickery, B.J. and Basu, R.I. Response of reinforced concrete chimneys to be vortex shedding. *Engineering Structures* **6**, 324–333, 1984.

Virlogeux, M. Cable vibrations in cable-stayed bridges. In *Bridge Aerodynamics*, Larsen, A. and Esdahl, S. (editors), Balkema, Rotterdam, 213–234, 1998.

Walshe, D.E. and Wootton, L.R. Preventing wind-induced oscillations of structures of circular section. *Proceedings of the Institution of Civil Engineers* **47**, 1–24, 1970.

Wootton, L.R. The oscillations of large circular stacks in wind. *Proceedings of the Institution of Civil Engineers* **43**, 573–598, 1969.

Warburton, G. *Reduction of Vibrations. Third Mallet–Milne Lecture*. John Wiley, Chichester, 1992.

Whitman, R.W. Analysis of foundation vibrations. In *Vibrations in Civil Engineering*, Skipp, B.O. (editor), Butterworth, London, 1966.

Wolf, J.P. *Dynamic Soil–Structure Interaction*. Prentice-Hall, Englewood Cliffs, NJ, 1985.

Wolf, J.P. *Foundation Vibration Analysis Using Simple Physical Models*. Prentice-Hall, Englewood Cliffs, NJ, 1994.

Wyatt, T.A. *Proceedings of a Symposium on the Dynamic Behaviour of Bridges*. TRRL Supplementary report 275 (paper 2). TRRL, London, 1977.

Wyatt, T.A. *Evaluation of Gust Response in Practice: Wind Engineering in the Eighties*. CIRIA, London, 1981, Chapter 7.

Wyatt, T.A. An assessment of the sensitivity of lattice towers to fatigue induced by wind gusts. *Engineering Structures* **6**, 262–267, 1984.

Wyatt, T.A. and Best, G. Case study of the response of a medium-height building to wind-gust loading. *Engineering Structures* **6**, 256–261, 1984.

Wyatt, T.A. and Scruton, C. *A Brief Survey of the Aerodynamic Problems of Bridges, Bridge Aerodynamics*. TTL, London, 1981, Chapter 1.

Yu, J. and Jones, N. Numerical simulation of impact loaded steel beams and the failure criteria. *International Journal of Solids and Structures*. **34**, 3977–4004, 1997.